The Codex 5 Model:

Describing the Origin & Nature of
Dark Energy & Dark Matter

and

Predicting the Distribution of Energy / Matter in the Universe

(Updated 2014)

by

Fritz Hoffecker

The Codex 5 Model: Describing the Nature of Dark Energy and Predicting the Distribution of Energy/Matter in the Universe

Copyright 2008, 2014 Frank S. (Fritz) Hoffecker

All rights reserved. No part of this paper, including text, tables and graphics, may be reproduced or transmitted in any form, by any means (electronic, photocopying, recording, or otherwise) without prior written permission of the author.

CONTENTS

Part 1: Summary/Abstract 1
   Key Predictions about Dark matter and Dark Energy 1
   Predictions about the Death and Rebirth of the Universe 3
   Appendices: Other Implications of the Codex 5 Model 3
   The Approach Used in the Codex 5 Model 3
1.0   Background: Definitions, Hypothesis, Statements 6
2.0   Properties of {x/y} 8
   2.1 The Basic Pair and other Entities as Anapoles 8
   2.2 Guidelines on this Paper's Notations 10
3.0   Possible Relationships between Sets (or "Basic Pairs") {x/y} and {a/b} 12
   3.1   Initial State 1 (IS_1) and Next State 1 (NS_1) 13
   3.2   Initial State 2 (IS_2) and Next State 2 (NS_2) 15
   3.3   Initial State 3 (IS_3) and Next State 3 (NS_3) 18
   3.4   Initial State 4 (IS_4) and Next State 4 (NS_4) 20
   3.5   Percentage of Occurrences - Phases 21
      3.5.1 Interactions that Follow the Initial Situation – Phase 1 23
      3.5.2 Interactions that Follow Phase 1: Phase 2 25
      3.5.3 Interactions that Follow Phases 1 & 2: Phase 3 28
      3.5.4 Interactions that Follow Phases 1, 2 & 3: Phase 4 30
4.0   Part 1 Conclusions 33
Part 2: Death and Rebirth of the Universe - Summary/Abstract 35
Appendix 1 – Codex 5: Implications 50
Appendix 2 – Other Predictions and Comments based on the Codex 5 Model ... 66

Part 1: Summary/Abstract

Key Predictions about Dark matter and Dark Energy

The purpose of Part 1 of this paper is to present the Codex 5 Model and associated reasoning, which:

1) Propose the existence of the most fundamental type of Energy / Matter, herein referred to as the "Basic Pair";
2) Describe the Nature of Dark Energy and Dark Matter; and
3) Predict the Distribution of Energy / Matter in the Universe.

Post-Big Bang, the Codex 5 Model predicts the following distribution of the types of matter in the universe, as follows:

Dark Energy: 71.43%
Dark Matter: 23.93%
Baryonic Matter: 4.64%

As shown in Table 1 below, note also that the Model predicts two different types of Dark Matter, which I call "Light" and "Heavy" for reasons described later in this paper.

Table 1: Distribution of Energy / Matter Types in the Universe

Final Distribution	% of Total E/M in each type of Entity
Dark Energy	71.43%
"Light" Dark Matter	21.43%
"Heavy" Dark Matter	2.50%
Baryonic Matter	4.64%
Total	100.00%

Current technology is unable to completely verify these predictions, but they are in line with other measurements such as those from NASA's Wilkinson Microwave Anisotropy Probe (WMAP):

WMAP Categories	% of Total E/M in each Category
Dark Energy	~71.4%
Dark Matter	~24.0%
Baryonic Matter	~4.6%

Table 1a below combines the Codex 5 Model's predictions with WMAP's observations:

Table 1a: Distribution of Energy / Matter Types in the Universe

Codex 5 Model: Final Distribution	Codex 5 Model: % of Total Energy / Mass in each type of Entity	WMAP Category	WMAP % distribution	Variation: Gross Diff: "xx%"
Dark Energy	71.43%	Dark Energy	71.4%	Gross Diff: 0.03%
"Light" Dark Matter	21.43%			
"Heavy" Dark Matter	2.50%			
Total Dark Matter	23.93%	Dark Matter	24.0%	Gross Diff: 0.07%
Baryonic Matter	4.64%	Baryonic Matter	4.6%	Gross Diff: 0.04%
Total	100.00%		100%	

Note: WMAP figures come from: http://map.gsfc.nasa.gov/universe/uni_matter.html (as of Nov 2013).

As shown in Table 1a, the Codex 5 Model's predictions are very close to WMAP data. These tight correlations support the following:
1) The existence of the "Basic Pair" as the most fundamental type of Energy / Matter.
2) The rationale in this paper.

3) The associated Model of Energy / Matter properties, interactions and evolution.
4) The hypotheses presented in Part 2 of this paper ref the Big Bang, and the subsequent demise and rebirth of the Universe.

Predictions about the Death and Rebirth of the Universe

Note that in this paper, **Part 2: Death and Rebirth of the Universe**, shows that the Model described in Sections 1.0 through 4.0 (below) has implications for how our current Universe started, how it will decline and end, and how a new Universe will be recreated. Figures 2-1 through 2-15 and the accompanying text show how the Codex 5 Model is integrated with: 1) the expanding Universe, 2) descriptions of subsequent phases of implosion, and 3) another Big Bang.

Appendices: Other Implications of the Codex 5 Model

Appendix 1 - Codex 5: Implications presents additional hypotheses that the Codex 5 Model leads to. These hypotheses address the following questions:

1) Dark Energy: What does it consist of? Where did it come from?
   Dark Matter: What does it consist of? Where did it come from?
2) How were elementary particles created? The Codex 5 Model describes this process at the lowest existing energy levels.
3) Density of Vacuum: For the inflation of the Universe after the Big Bang, a surrounding non-zero vacuum expectation value (VEV) is required. How does this non-zero VEV come about?
4) Inflation: Why and how does matter/energy 'expand' from the initial singularity of the Big Bang?
5) Fluctuations: WMAP has detected early fluctuations in the Cosmic Microwave Background (CMB) and in the Cosmic Neutrino Background (CNB)? What causes these fluctuations?
6) Cosmological Constant: What is Λ?: $\Lambda = 0$, or $\Lambda > 0$, or $\Lambda < 0$?
7) "The Edge of the Universe": The Universe is expanding, so what is on the "outer edge"? Where is it expanding to?
8) The Singularity: What caused the Singularity where the Big Bang occurred?

Appendix 2 – Other Predictions and Comments based on the Codex 5 Model speaks against String Theory, Branes, and Multiverses, and presents a few examples of evidence supporting the existence of Dark Matter and Dark Energy.

The Approach Used in the Codex 5 Model

In attempting to understand and describe physical reality, the Codex 5 Model uses an approach that is similar to Dynamical Systems and Cellular Automata. In summary:

The Model starts with very simple entities (i.e., Basic Pairs), applies rules to the Entities, and determines what happens at each iteration of the rules.

The rules in the Model show a Dynamical System (i.e., our Universe) where the simplest of entities (Basic Pairs) are subject to certain rules, and – as these rules guide the behavior of these entities - eventually evolve into the actual Universe we see now. Stephen Wolfram wrote in *A New Kind of Science*: "…it also means that if one once discovers a rule that reproduces sufficiently many features of the universe, then it becomes extremely likely that this rule is indeed the final and correct one for the whole universe." (page 469) Weinberg also wrote: "To find the behavior of the universe one potentially needs to know not only its rule but also its initial condition. Like the rule, I suspect that the initial conditions will turn out to be simple." (page 1026)

The Codex 5 Model aligns with Mr. Weinberg's beliefs about the simplicity of the Initial State, and the importance of "the rules": The Codex 5 Model is an example of this approach:

- The Model maintains that the Universe started with simple conditions (e.g., simple entities, Basic Pairs).
- It applies simple rules.
- It predicts what WMAP has measured as key features of the current Universe, e.g., the creation and percentages of the Universe's Energy/Matter that are made of Dark Energy, Dark Matter, and Baryonic Matter.

Credence is leant to the Codex 5 Model because it predicts current percentages of Dark Matter and Dark Energy that are nearly identical to the percentages that WMAP has observed and measured.

What this Work is and isn't

I have a M.S. degree and have written a number of "academic" papers in the past, but I'm not a member of the academy and this work is obviously not a true academic paper.

This might lead some people to discount the Codex 5 Model, but I decided on a more "discursive," open, non-academic approach because it better suits my objectives, which are:

- To present a new model that is substantially different from other approaches to understanding the origins, nature and ultimate fate of Dark Matter and Dark Energy.
- To show that using the principals of Cellular Automata and Dynamical Systems can lead to a model that accurately predicts the percentages of Dark Energy, Dark Matter, and Baryonic Matter in the current Universe.
- To show that this model also indicates a) what happened at the most recent Big Bang, and what will happen and b) what the future of our current Universe will be.
- To show that the Codex 5 Model has implications for other interesting, related subjects (see Appendices).

- To lay out a model and framework that can guide further related inquiries into the nature of Dark Matter and Dark Energy, plus other areas such as views of the Big Bang and the fate of the current iteration of the Universe.

I understand that members of "The Academy" could easily criticize this work, but I suggest that they first review and try to understand the Codex 5 Model. For example, it does accurately predict how energy/matter is distributed in the Universe.

1.0 Background: Definitions, Hypothesis, Statements

<u>Definitions</u>

For the purposes of this paper, I'm using the following definitions:

Absolute-Nil is defined as a "Basic Entity" that is "complete Nothingness". For example, it is not composed of baryonic matter, does not emit or reflect any energy in the electro-magnetic spectrum. Figuratively, it can be thought of as the tiniest possible bit of nothingness in the Universe.

Non-Nil is defined as a "Basic Entity" that consists of matter that cannot be further decomposed. Figuratively, it can be thought of as the tiniest possible bit of matter in the Universe.

Absolute-Nil and Non-Nil can exist only together as a single entity, which the Model calls a Basic Pair.

In this paper:

Non-Nil = x (or a)

That is, x (or a) represents the Non-Nil Basic Entity that is the smallest instantiation of matter in the Universe.

Absolute-Nil = y (or b)

That is, y (or b) represents the Absolute Nil Basic Entity that is the smallest instantiation of 'nothingness' in the Universe.

In these definitions, the term "small" or "smallest" refers to the amount of Energy that the Basic Entity possesses, and/or that can react with another Basic Entity's Energy.

<u>Statements / Hypothesis:</u>

For any set $\{x, y\}$,

1) This set is defined as a "Basic Pair".

2) Each x (a Basic Entity that is an instantiation of Non-Nil) is inextricably paired with a y (a Basic Entity that is an instantiation of Absolute Nil).
This set is written: $\{x/y\}$

In other words: $f(\{x,y\}) = \{x/y\}$

3) Additional Basic Pairs:
   Other sets can be formed as pairs a and b, where
   a = another Basic Entity – different from x - that is an instantiation of Non-Nil, and
   b = another Basic Entity – different from y - that is an instantiation of Absolute Nil.

   This set (or "Basic Pair") is written: {a/b}.

4) x and a have exactly the same properties. Each represents an instantiation of Non-Nil, and all Non-Nils have the same properties.
y and b have exactly the same properties. Each represents an instantiation of Absolute-Nil, and all Absolute -Nils have the same properties.

2.0 Properties of {x/y}

In discussing what a Basic Pair (BP) is, it is difficult to make drawings that truly represent the fact that the BP isn't really "that round thing in the diagram." The drawings are simply a shorthand method used for reference as I discuss what a BP is, what is does, how it acts in different situations.

2.1 The Basic Pair and other Entities as Anapoles

Having said that, I think it's useful to point out an important characteristic of the BP (and of related entities, which I'll get to eventually). Again, I'll resort to diagrams to make the following point: The BP does not revolve or spin around a North-South axis, and therefore does not have a North or South Pole. This fact probably explains why Dark Matter and Dark Energy are so hard to detect with traditional tools and instrumentation that are designed to detect and measure electro-magnetic phenomena. BPs and related Entities are, in fact, "anapoles."

Diagram 1 below shows the typical idea of a BP that I'll use below in describing their interactions:

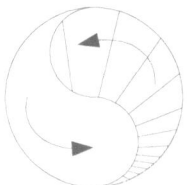

Diagram 1: "Typical" Basic Pair (BP)

Diagram 2 below picks an arbitrary point (represented by the "x") on any BP, and shows how it moves as the BP entity revolves. Note that the rotation of the overall entity is what we'd call "counter-clockwise," but that the point's movement indicates that the BP is not rotating strictly around an axis. Instead, the BP is "tumbling," such that it does not rotate around a specific axis: This makes it an anapole.

Diagram 2: The Basic Pair is an Anapole

2.2 Guidelines on this Paper's Notations

For any set {x/y}, no energy is perceptible to current methods of observation. Figure 1 presents a 2-D representation of what is a 3-D state in reality:

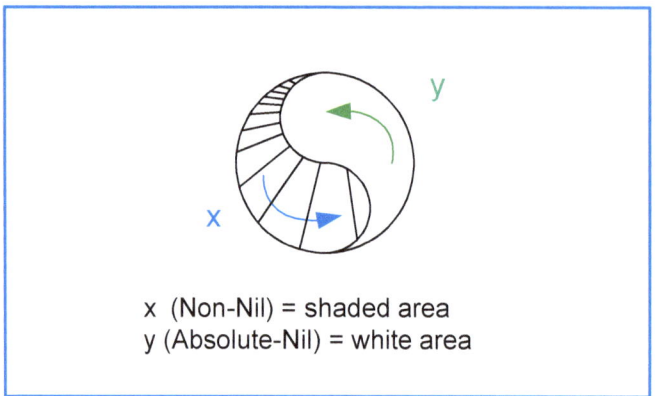

x (Non-Nil) = shaded area
y (Absolute-Nil) = white area

Figure 1 - {x $_{->}$ / $^{<-}$ y} = A Balanced State

The blue and green arrows in the diagram indicate the state of balance inherent in {x/y}.

In this paper, the set represented in this diagram will be written {x -> / <- y}, but with the following additional notations:

1) For Figure 1 above, where the "head" (or larger part) of x is at the bottom of the diagram and "headed" toward the right (per the direction of the arrow in Figure 1), x will be written as:

x $_{->}$

Note that the blue arrow is subscripted to show the location of x's "head."

In summary, this indicates that the "larger" section (the "head") of Non-Nil (x) is at the bottom of the diagram and that this "larger" section of Non-Nil always moves toward the "smaller" section (the "tail") of Absolute-Nil (y).

2) For Figure 1 above, where the "head" of y is at the top of the diagram, and is "headed" toward the left, y will be written as:

$^{<-}$ y

Note that the green arrow is super-scripted to show the location of y's "head" (see Figure 1).

In summary, the "larger" section of Absolute -Nil (the "top" of y in Figure 1) is always "headed" (per the direction of the green arrow) toward the "smaller" section of Non-Nil (the "top" of x in Figure 1).

Statement / Hypothesis:
For {x -> / <- y} , the properties of x and y are as follows:

Energy of x = Energy of y, or

E (x) = E (y)

For the set {x/y}, following is a Statement of properties:

Property 1: x (Non-Nil) is inseparable from y (Absolute-Nil).
Conversely,
A property of y (Absolute-Nil) is that it is inseparable from x (Non-Nil).

This can be expressed as:

x ⇄ y

Property 2: Energy F(x) -> y = Energy F(y) -> x

This means:
Because the two Energies of x and y are equal, the Set (or "Basic Pair") {x -> / <- y} is said to be in balance.
x continues to be inseparable from y, and y continues to be inseparable from x.
x and y exert the same amount of Force on each other.

3.0 Possible Relationships between Sets (or "Basic Pairs") {x/y} and {a/b}

"...it also means that if one once discovers a rule that reproduces sufficiently many features of the universe, then it becomes extremely likely that this rule is indeed the final and correct one for the whole universe...To find the behavior of the universe one potentially needs to know not only its rule but also its initial condition. Like the rule, I suspect that the initial conditions will turn out to be simple." Stephen Wolfram, A New Kind of Science, pages 469, 1026.

This section describes the relationships that can arise between two separate sets, {x/y} and {a/b}.

1) Initial State
For any two different sets, {x/y} and {a/b}:
An Initial State will be described.

An "Initial State" is defined as the relationship between the two different sets at any specific instant of measurement/observation (whether this be a "real" measurement or observation made by some person and/or machine, or a hypothetical measurement/observation as described herein).

2) Force Field Intersection State
A property of each Basic Pair - e.g., {x/y} and {a/b} – is that it has some Energy. (Since Non-Nil has been defined as "matter," it has Mass, which in turn indicates Energy. Though Absolute-Nil has no Mass, it exerts force, i.e., has energy).
The degree that this Energy extends (i.e., herein referred to as the "Force Field") is defined as the edge of the circle on the diagram.
The Force Field Intersection State occurs when the Force Field extent of one Basic Pair ({x/y} in these examples) intersects or "touches upon" the Force Field of another Basic Pair ({a/b} in these examples).

3) Next State
Following the "Force Field Intersection State" will be a "Next State", which is defined as any subsequent state of that Relationship between the Basic Pairs {x/y} and {a/b}. The term "subsequent" is used to place these perceptions in the context of a forward-arrow of Time. The starting point of this forward-arrow is defined at the instant when the Big Bang expansion begins.

Several "Initial States" and "Next States" are described below. The "Initial States" are defined as:
- Initial State 1 (IS$_1$)
- Initial State 2 (IS$_2$)
- Initial State 3 (IS$_3$)
- Initial State 4 (IS$_4$)

Force Field Interaction states are shown on the diagrams below, but don't require specific designations (e.g., "FI$_1$" or something similar).

"Next States" are designated:
- Next State 1 (NS$_1$), i.e., the State that follows Initial State 1 (IS$_1$)
- Next State 2 (NS$_2$), i.e., the State that follows Initial State 2 (IS$_2$)
- Next State 3 (NS$_3$), i.e., the State that follows Initial State 3 (IS$_3$)
- Next State 4 (NS$_4$), i.e., the State that follows Initial State 4 (IS$_4$)

3.1   Initial State 1 (IS$_1$) and Next State 1 (NS$_1$)

Initial State 1 (IS$_1$) is shown is Figure 2 below.

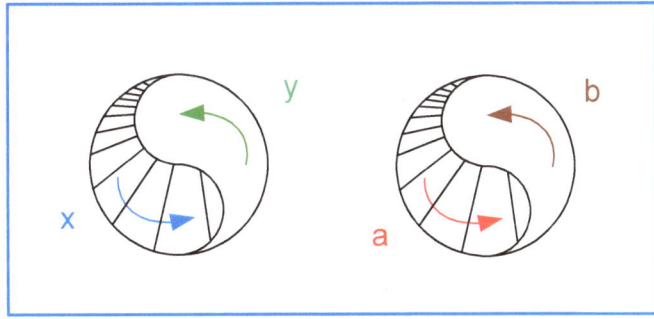

Figure 2 – Initial State 1 (IS$_1$)

In Initial State 1, $\{x_\rightarrow / \leftarrow y\}$ and $\{a_\rightarrow / \leftarrow b\}$ have the same properties (e.g., the amount of Energy and the topology in each Basic Pair), as indicated by their identical depictions in Figure 2.

Figure 3 shows:
1) The Initial State
2) The Intersection of the Force Fields of the Entities represented by Sets $\{x_\rightarrow / \leftarrow y\}$ and $\{a_\rightarrow / \leftarrow b\}$
3) The Next State (NS$_1$)

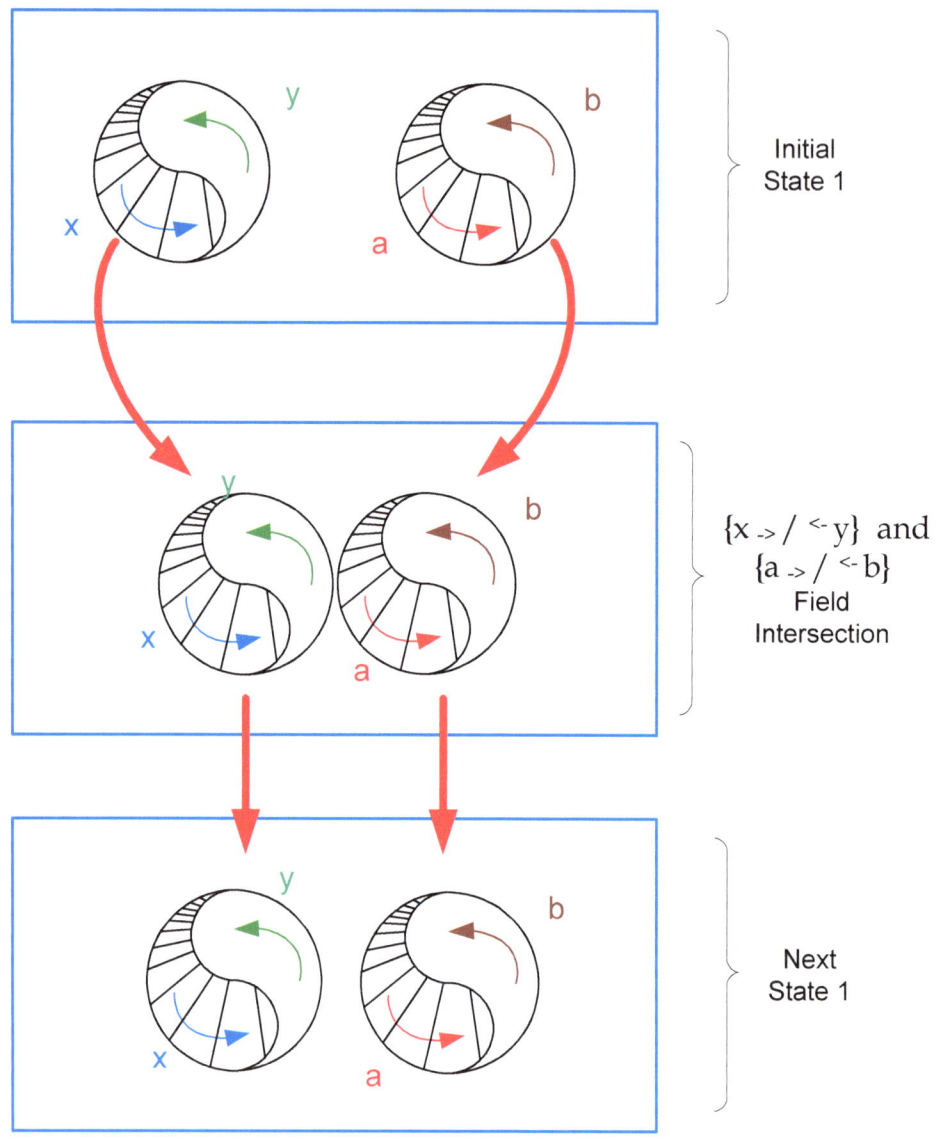

Figure 3 – IS_1 => NS_1

This "transition" is expressed as follows:

    State IS_1 (which is {x -> / <- y} and {a -> / <- b}
=> State NS_1 (which is also {x -> / <- y} and {a -> / <- b})
Where "=>" means "leads to"

Next State 1 is expressed as follows:

NS$_1$ = {x $_\to$ / $^\leftarrow$ y} || {a $_\to$ / $^\leftarrow$ b}

Where "||" means that the two Basic Pairs "stay separate from each other."

In other words, this means that {x/y} and {a/b} remain the same relative to each other, i.e., stay separate from each other, or:

Initial State 1 = Next State 1

3.2 Initial State 2 (IS$_2$) and Next State 2 (NS$_2$)

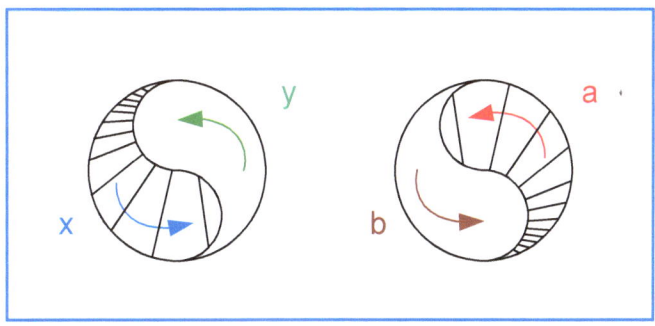

Figure 3 – Initial State 2 (IS$_2$)

In Initial State 2, the set {a/b} (on the right in Figure 3) is different than it is in Initial State 1. As shown in the diagram, a and b have reversed positions from where they were in Initial State 1.

To show this difference, the sets are written

not as: {x $_\to$ / $^\leftarrow$ y} and {a $_\to$ / $^\leftarrow$ b} - Initial State 1
but as: {x $_\to$ / $^\leftarrow$ y} and {b $_\to$ / $^\leftarrow$ a} - Initial State 2

Figure 4 shows the sequence starting from Initial State 2:

1) Initial State 2 (IS$_2$)
2) Force Field Interaction between the two Basic Entities represented as Sets {x $_\to$ / $^\leftarrow$ y} and {b $_\to$ / $^\leftarrow$ a}
3) Next State 2 (NS$_2$).

IS$_2$ (expressed as {x $_{\rightarrow}$ / $^{\leftarrow}$ y} and {b $_{\rightarrow}$ / $^{\leftarrow}$ a} => {xa $_{\rightarrow\rightarrow}$ / $^{\leftarrow\leftarrow}$ yb}
Where "=>" means "leads to"

NS$_2$ = | ->{xa $_{\rightarrow\rightarrow}$ / $^{\leftarrow\leftarrow}$ yb} <- |

Where " | ->" indicates that {x/y} <u>has merged with</u> {a/b} and
"<- | " means that {a/b} <u>has merged with</u> {x/y}
The 'size' of the Entity created (called a Large Pair) in NS$_2$ is shown as:

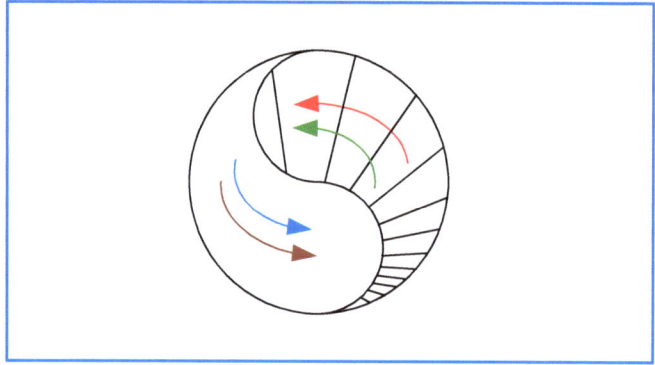

Figure 4 – NS$_2$ or {xy/ab}

Figure 5 below shows the transition from IS$_2$ to NS$_2$.

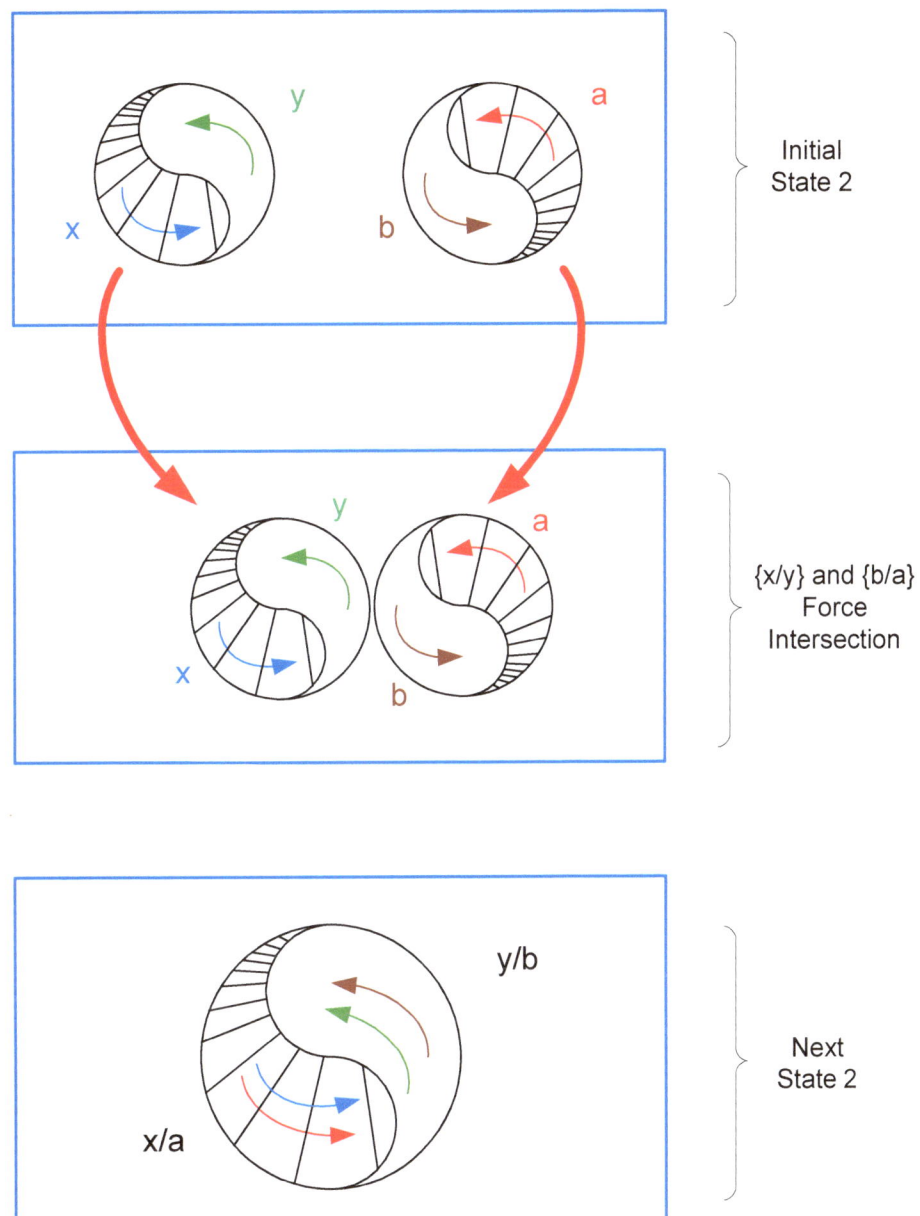

Figure 5 – Transition from IS$_2$ to NS$_2$

Note that the Energy in NS$_2$ is the sum of the Energy of {x/y} and Energy of {a/b} This is expressed as:

Energy F({xa/yb} = Energy ({x/y}) + Energy ({a/b})

Note also that NS$_2$ is in balance, which is expressed as:

Energy F(x/a) = Energy F(y/b)

Because the two Energies are equal, NS$_2$ is said to be in balance. x/y is attracted to y/b in the same proportion that y/b is attracted to x/a.

However, note that a new Pair, {xa/yb}, has been created. From an Energy point of view it is larger than each of the Basic Pairs, {x/y} and {a/b}.

3.3   Initial State 3 (IS$_3$) and Next State 3 (NS$_3$)

In Initial State 3, the relative positions of a and b in the set {a/b} are different than in Initial States 1 and 2 (see Figure 6).

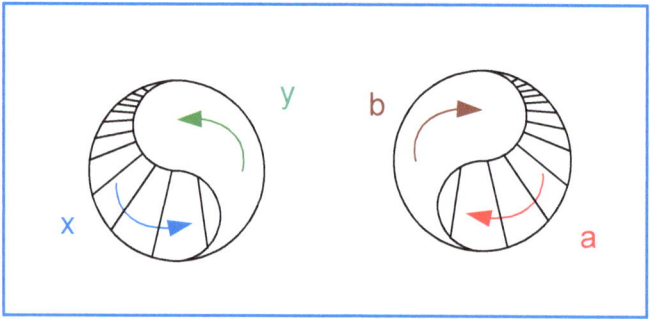

Figure 6 – Initial State 3

IS3 is written: {x -> / <- y} and {b -> / <- a}

For Initial State 3, in the "Next State", designated NS$_3$, {x -> / <- y} and {b -> / <- a} <u>stay separate</u>, and it is possible that they become farther apart, i.e., each repulses the other. This is written as:

NS$_3$ = {x -> / <- y} <-| |-> {b -> / <- a}

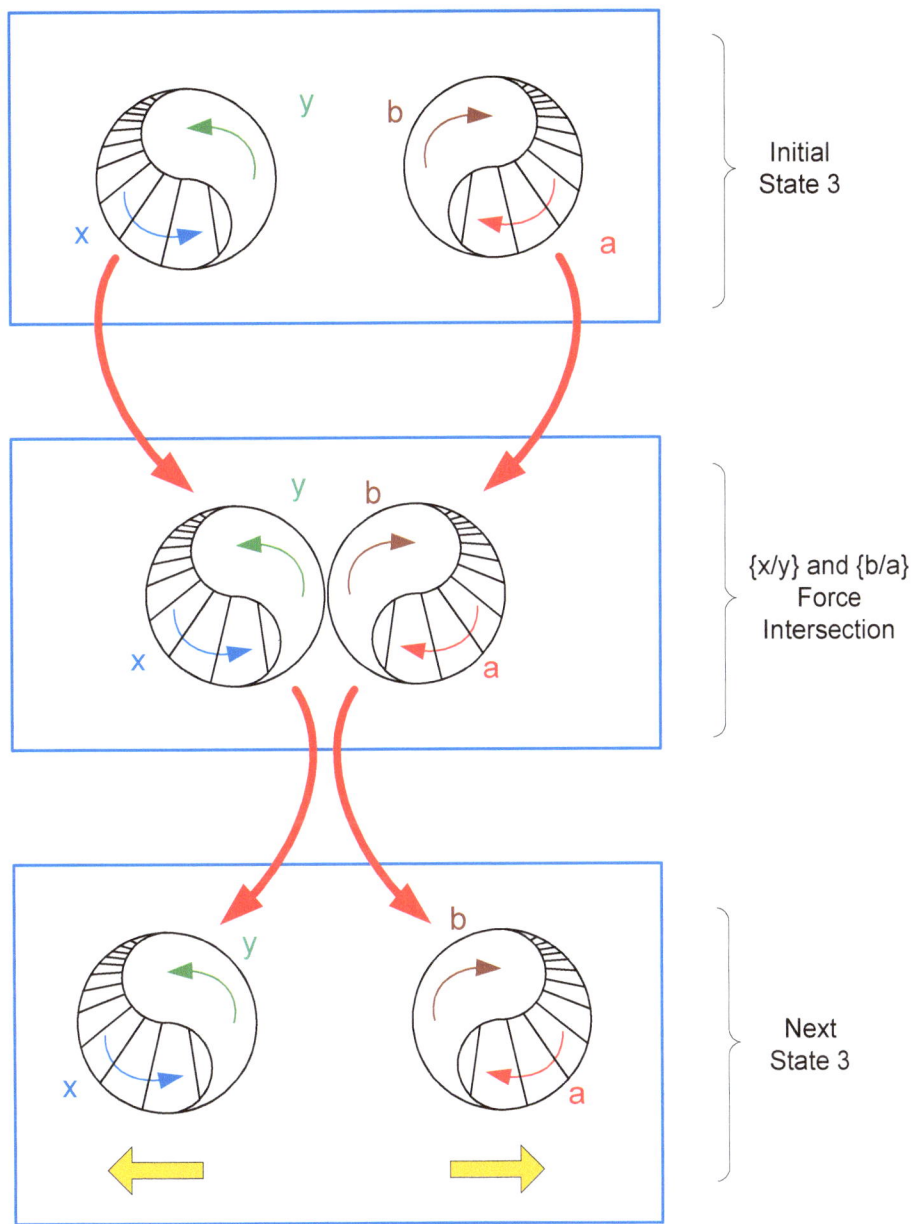

Figure 7 – Transition from IS_3 to NS_3

3.4 Initial State 4 (IS$_4$) and Next State 4 (NS$_4$)

In Initial State 4, the relative positions of a and b in the set {a/b} are different than in States 1, 2 and 3 (see Figure 8).

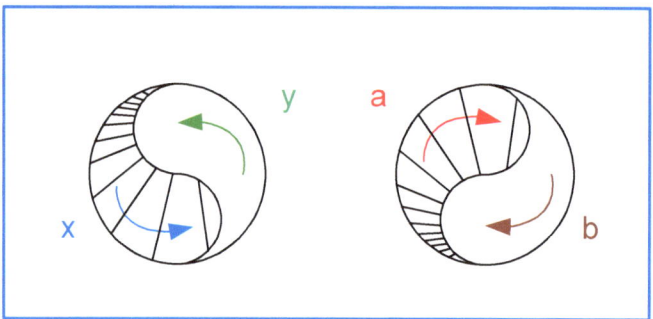

Figure 8 - *Initial State 4*

For *Initial State 4*, in the "Next State 4", {x/y} and {a/b} <u>stay separate</u>. This is written as:

Next State 4 = {x/y} <-| |-> {a/b}

See Figure 9 below for a graphical representation of the transition from Initial State 4 to Next State 4.

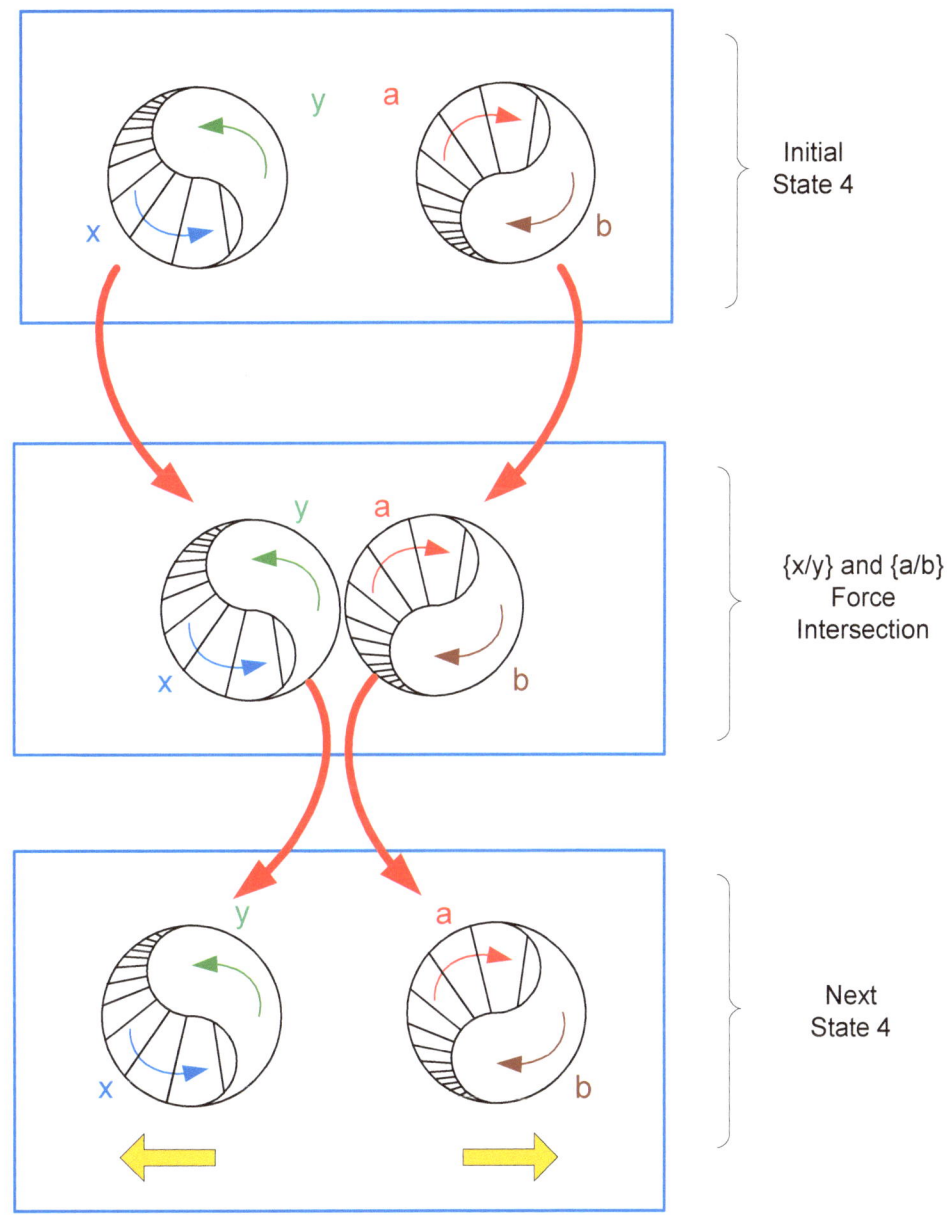

Figure 9 – Transition from IS₄ to NS₄

3.5 Percentage of Occurrences - Phases

As shown above in Section 3.1 through 3.4, Initial States 1 through 4 lead to Next States 1 through 4.

In Sections 3.5.1 through 3.5.4, I'll consider further implications.

Initial Conditions
This analysis starts by looking at a situation that includes the following Basic Pairs (BPs):

BP 1 and BP1b, where BP1b has the same characteristics and properties of BP1
BP 2 and BP2b, where BP2b has the same characteristics and properties of BP2
BP 3 and BP3b, where BP3b has the same characteristics and properties of BP3
BP 4 and BP4b, where BP4b has the same characteristics and properties of BP4

In summary:

BP1 and BP1b look like this:

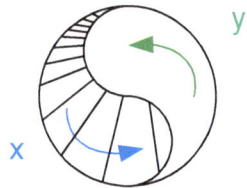

BP2 and BP2b look like this:

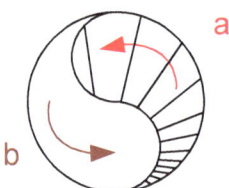

BP3 and BP3b look like this:

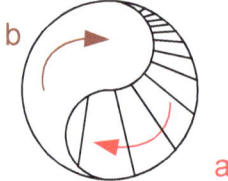

BP4 and BP4b look like this:

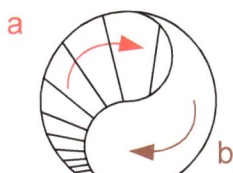

3.5.1 Interactions that Follow the Initial Situation – Phase 1

In this section, the analysis considers all the possible interactions between the eight Basic pairs.

The following equation shows the total number of possible interactions (x) between any "n" number of separate items:

$$x = \frac{n(n-1)}{2}$$

In this case, where I'm considering 8 separate Basic Pairs, n = 8,
So, using the formula above: x = 28. In fact, in situ, there are some unknown number of Basic Pairs (not just 8) – e.g., many billions and billions – but these 8 represent all the possible sets of interaction that are possible

Table 2 below shows the 28 different interactions, and the results of each interaction. The possible results (as shown above in Sections 3.1 through 3.5) are either:
1) The two Basic Pairs that are interacting <u>stay separate</u>. That is, in the "Next State" (see Sections 3.1 through 3.4), the BPs stay separate, OR
2) The two basic Pairs <u>merge into a single "larger"</u> Basic Pair (see Section 3.2).

Table 2: Results of 28 Possible Interactions

1st Basic Pair	2nd Basic Pair	Result of Interaction
1	2	BPs merge
1	3	BPs stay separate
1	4	BPs stay separate
1	1b	BPs stay separate
1	2b	BPs merge
1	3b	BPs stay separate
1	4b	BPs stay separate
2	3	BPs stay separate
2	4	BPs stay separate
2	1b	BPs merge
2	2b	BPs stay separate
2	3b	BPs stay separate
2	4b	BPs stay separate
3	4	BPs merge
3	1b	BPs stay separate
3	2b	BPs stay separate
3	3b	BPs stay separate
3	4b	BPs merge
4	1b	BPs stay separate
4	2b	BPs stay separate
4	3b	BPs merge
4	4b	BPs stay separate
1b	2b	BPs merge
1b	3b	BPs stay separate
1b	4b	BPs stay separate
2b	3b	BPs stay separate
2b	4b	BPs stay separate
3b	4b	BPs merge

Table 3 below shows:
1) The percentage of interactions that result in the Basic Pairs staying separate and
2) The percentage of interactions that result in the Basic Pairs merging into a "larger" Basic Pair (which will be called Large Pairs, or "LgP").

Table 3: Phase 1 - Results of 28 Possible Interactions

Phase 1	Quantity	Percentage
Total # Interactions	28	
Results of Interactions:		
Basic Pairs (BPs) merge	8	28.57%
Basic Pairs (BPs) stay separate	20	71.43%

These percentages indicate that 71.43% of the Basic Pairs will never merge and never be detectable as baryonic matter, i.e., do not emit or reflect electromagnetic radiation that we can detect with current technology.

3.5.2 Interactions that Follow Phase 1: Phase 2

After the 28 initial interactions, 71.43% of the initial Basic Pairs (BPs) will remain BPs forever.
The remaining 28.57% of the initial BPs have merged into a total of 8 Large Pairs (LgPs).

In Phase 2, it is assumed that there are no possible interactions between BPs and LgPs. Only LgPs interact with other LgPs.
Table 4 below shows the possible interactions between the remaining 8 LgPs, and the results of each possible interaction:

Table 4: Phase 2 Interactions and Results

Interaction		Result
LgP1	LgP1b	stay separate
LgP2	LgP2b	stay separate
LgP3	LgP3b	stay separate
LgP4	LgP4b	stay separate
LgP1	LgP4b	stay separate
LgP2	LgP1b	merge
LgP3	LgP2b	stay separate
LgP4	LgP3b	merge
LgP1	LgP3b	stay separate
LgP2	LgP4b	stay separate
LgP3	LgP1b	stay separate
LgP4	LgP2b	stay separate
LgP1	LgP2b	merge
LgP2	LgP3b	stay separate
LgP3	LgP4b	merge
LgP4	LgP1b	stay separate

After Phase 2, the Energy/Matter of the universe is distributed as shown in Table 5 below:

Table 5 – Results after Phase 2

Phase 2	Type of Entity	Quantity of each Entity	Units of Energy / Matter (E/M) per Entity	Total Units of E/M per Entity Type		End of Phase 1: % of Total E/M in each type of Entity	End of Phase 2: % of Total E/M in each type of Entity
Throughout Phase 2, 71.4% of E/M remains in Basic Pairs	BPs	40 -40 BPs remain after Phase 2	1 -Each BP has 1 Unit of E/M	40 -BPs make up total of 40 units of E/M		71.43%	71.4%
So after Phase 2, 21.4% of all E/M is now in Large Pairs (LgPs)	LgPs	6 (pro-rated to stay at E/M ceiling of 56) -6 LgPs remain after Phase 2	2 -Each LgP has 2 Units of E/M	12 -LgPs make up total of 12 units of E/M	12 of 16 E/M units (75%) are in LgPs	28.57% - is the percent of total energy that's not in BPs	21.43% - (0.75 * .286): means after Ph 2, 21.4% of total E/M is in LgPs
Remaining 7.1% of E/M is in Very Large Quads (VLQs)	VLgQs	1 (pro-rated to stay at E/M ceiling of 56) been created, & remains after Phase 2	4 -Each VLgQ has 1 Units of E/M	4 -VLQs make up total of 4 units of E/M	4 of 16 E/M units (25%) are in VLQs		7.14% - (0.25 * .286): means after Ph 2, 7.14% of total E/M is in VLQs
				Note total of 56 units of E/M			

3.5.3 Interactions that Follow Phases 1 & 2: Phase 3

In Phase 3, it is assumed that there are no possible interactions between BPs and any other entities (i.e., LgPs, VLQs). The only interactions will be:

1) LgPs with other LgPs
2) LgPs with VLQs
3) VLQs with other VLQs

Table 6 below shows the possible interactions among the LgPs and VLQs (i.e., the ones in existence after Phase 2), and the results of each possible interaction:

Table 6 – Interactions that Occur in Phase 3, and Results

Start of Phase 3		Result of each interaction	End of Phase 3, Start of Phase 4	
VLQ-1b	VLQ-3b	sep =>	VLQ-1b	VLQ-3b
VLQ-2b	VLQ-4b	sep =>	VLQ-2b	VLQ-4b
VLQ-1b	VLQ-4b	sep =>	VLQ-1b	VLQ-4b
VLQ-2b	VLQ-3b	sep =>	VLQ-2b	VLQ-3b
VLQ-1b	VLQ-2b	merge =>	VLOct-2b	
VLQ-3b	VLQ-4b	merge =>	VLOct-3b	

After Phase 3, the Energy/Matter of the universe is distributed as shown in Table 7 below:

Table 7 – Results after Phase 3
(Note: grayed out Phase 2 results are for reference only)

Phase		Type of Entity	Quantity of each Entity	Energy / Mass per Entity	Total Units of Energy per Entity Type		End of Phase 1: % of Total E/M in each type of Entity	End of Phase 2: % of Total E/M in each type of Entity	End of Phase 3: % of Total E/M in each type of Entity
Phase 3 results	Throughout Phase 3. 71.4% of E/M remains in Basic Pairs	BPs	40	1	40	71.4%	71.4%	71.4%	71.4%
	Throughout Phase 3. 21.4% of E/M remains in Lg Pairs	LgPs	6	2	12	21.4%	21.4%	21.43% - (0.75 * .286): means after Ph 2, 21.4% of total E/M is in LgPs	21.43% - (0.75 * .286): means after Ph 3, 21.4% of total E/M is in LgPs
Phase 2 results	So after Phase 2, 21.4% of all E/M is now in Large Pairs (LgPs)	LgPs	6 (pro-rated to stay at E/M ceiling of 56) -6 LgPs remain after Phase 2	2 -Each LgP has 2 Units of E/M	12 -LgPs make up total of 12 units of E/M	12 of 16 E/M units (75%) are in LgPs	28.57% - is the percent of total energy that's not in BPs	21.43% - (0.75 * .286): means after Ph 2, 21.4% of total E/M is in LgPs	21.43% - (0.75 * .286): means after Ph 3, 21.4% of total E/M is in LgPs
	Remaining 7.1% of E/M is in Very Large Quads (VLQs)	VLgQs	(pro-rated to stay at E/M ceiling of 56) -1 VLQ has been created, & remains after Phase 2	4 -Each VLgQ has 1 Units of E/M	4 -VLQs make up total of 4 units of E/M	4 of 16 E/M units (25%) are in VLQs		7.14% - (0.25 * .286): means after Ph 2, 7.14% of total E/M is in VLQs	
Phase 3 results	2/3 of VLQ's remain as VLQs	VLQs	8 = gross # 2.67 = prorated #	4	2.67	66.67%			4.76% (.67*7.1%): means after Phase 3, 4.76% of total E/M is in VLQs
	1/3 of VLQ's become VLOctets	VLO's	2 = gross # 1.33 = prorated #	8	1.33	33.33%			2.38% (.33 * 7.1%): means after Phase 3, 2.38% of total E/M is in VLOcts

3.5.4 Interactions that Follow Phases 1, 2 & 3: Phase 4

In Phase 4, it is assumed that there are:
1) No possible interactions between BPs and any other entities (i.e., LgPs, VLQs, VLOcts).
2) No possible interactions between LgPs and any other entities.

The only interactions will be:

1) VLQs with other VLQs
2) VLQs with other VLOcts
3) VLOcts with other VLOcts

Table 7 below shows the possible interactions among the VLQs and VLOcts (i.e., the ones in existence after Phase 3), and the results of each possible interaction:

Table 7 – Interactions that Occur in Phase 4

Start of Phase 4		Result of each interaction	End of Phase 4	
VLQ-1b	VLQ-3b	sep =>	VLQ-1b	VLQ-3b
VLQ-2b	VLQ-4b	sep =>	VLQ-2b	VLQ-4b
VLQ-1b	VLQ-2b	merge =>	VLOct-2b	
VLQ-1b	VLQ-4b	sep =>	VLQ-1b	VLQ-4b
VLQ-2b	VLQ-3b	sep =>	VLQ-2b	VLQ-3b
VLQ-3b	VLQ-4b	merge =>	VLOct-3b	
VLOct-2b	VLOct-3b	sep =>	VLOct-2b	VLOct-3b
VLQ-1b	VLOct-2b	merge =>	XL12-2b	
VLQ-2b	VLOct-3b	sep =>	VLQ-2b	VLOct-3b
VLQ-3b	VLOct-2b	sep =>	VLQ-3b	VLOct-2b
VLQ-4b	VLOct-3b	merge =>	XL12-3b	
VLQ-1b	VLOct-3b	sep =>	VLQ-1b	VLOct-3b
VLQ-2b	VLOct-2b	sep =>	VLQ-2b	VLOct-2b
VLQ-3b	VLOct-3b	sep =>	VLQ-3b	VLOct-3b
VLQ-4b	VLOct-2b	sep =>	VLQ-4b	VLOct-2b

After Phase 4, the Energy/Matter of the universe is distributed as shown in Table 8 below:

Table 8 - Results after Phase 4

Phase 4		Type of Entity	Quantity of each Entity	Units of Energy / Mass per Entity	Total Units of Energy per Entity Type		End of Phase 1: % of Total E/M in each type of Entity	End of Phase 2: % of Total E/M in each type of Entity	End of Phase 3: % of Total E/M in each type of Entity	End of Phase 4: % of Total E/M in each type of Entity
Phase 4 Results	Throughout Phase 4, 71.4% of E/M remains in Basic Pairs	BPs	40	1	40	71.4%	71.4%	71.4%	71.4%	71.4%
	Throughout Phase 4, 21.4% of E/M remains in Lg Pairs	LgPs	6	2	12	71.4%	21.4%	21.43% - (0.75 * 286): means after Ph 2, 21.4% of total E/M is in LgPs	21.43% - (0.75 * 286): means after Ph 3, 21.4% of total E/M is in LgPs	21.43%
Phase 4 Calcs						A = % of 7.14% in VLQ, VLOct or XL12:	B = E/M not in BPs and LgPs = 7.14%:	A x B		
Phase 4 Results	VLQ (each has 4 Units of E/M)	VLQs	14 = gross #	4		35.00%	7.14%	2.50%		2.50%
	VLOct (each has 8 Units of E/M)	VLOcts	10 = gross #	8		50.00%		3.57%		3.57%
	XL12 (each has 12 Units of E/M)	XL12s	2 = gross #	12		15.00%		1.07%		1.07%
								TOTAL ->		100.0%

Table 9 below provides a summary of Energy / Matter distribution in the universe after Phase 4:

Table 9 - Energy / Mass Distribution in the Universe

Final Distribution	Type of Entity	End of Phase 4: % of Total E/M in each type of Entity	
Throughout Phase 4, 71.4% of E/M remains in Basic Pairs	BPs	71.43%	Dark Energy
Throughout Phase 4, 21.4% of E/M remains in Large Pairs	LgPs	21.43%	"Light" Dark Matter
VLQ (each has 4 Units of E/M)	VLQs	2.50%	"Heavy" Dark Matter
VLOct (each has 8 Units of E/M)	VLOcts	3.57%	VLOct's & XL12's will eventually develop into Baryonic Matter
XL12 (each has 12 Units of E/M)	XL12s	1.07%	

4.0 Part 1 Conclusions

In conclusion, Part 1 of this paper has demonstrated and/or proposed the following:
1) The most fundamental instantiation or type of Energy / matter, is the Basic pair, as defined above.
2) Dark Energy consists of Basic Pairs.
3) Post-Big Bang, the interactions between the different types of Energy / Matter (E/M) lead to the following distribution of E/M in our current universe (see Table 9 above):

Dark Energy:	71.43%
Light Dark Matter:	21.43%
Heavy Dark Matter:	02.50%
Baryonic Matter:	04.64%

Figure 10 below provides a graphical representation of the evolution from the beginning of Phase 1 through the end of Phase 4.

Part 2 will address further implications of the CODEX 5 Model.

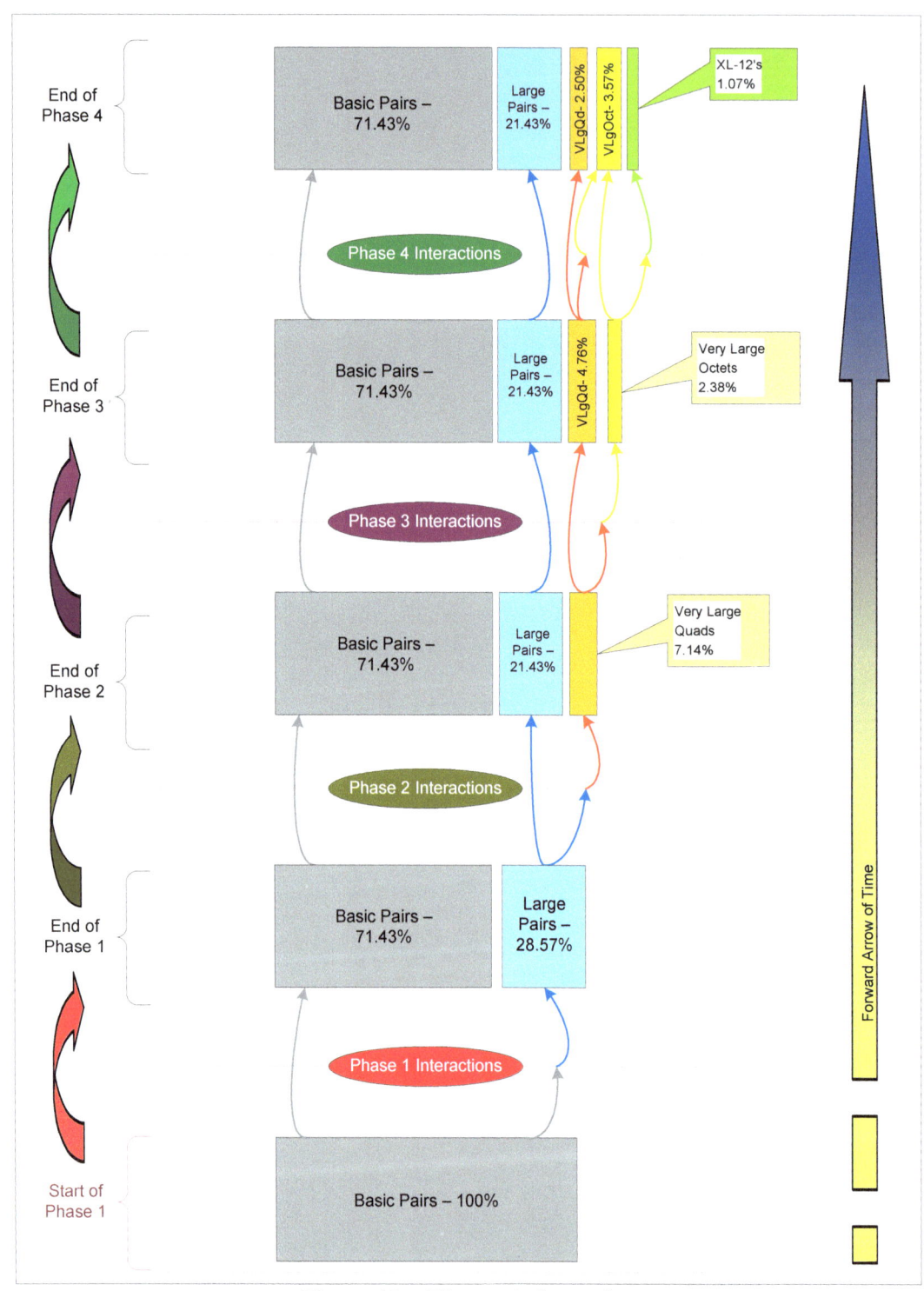

Figure 10 – Phases 1 through 4

Part 2: Death and Rebirth of the Universe - Summary/Abstract

The purpose of Part 2 of this paper is to propose a vision of how the current version of the Universe will end, and how a subsequent Universe will then begin.

The model described above in Sections 1.0 through 4.0 has implications for how our current Universe started, how it will decline and end, and how a new Universe will be recreated. Figures 2-1 through 2-15 below show how the Codex 5 Model is integrated with 1) the expanding Universe, 2) through subsequent phases of implosion and 3) another Big Bang.

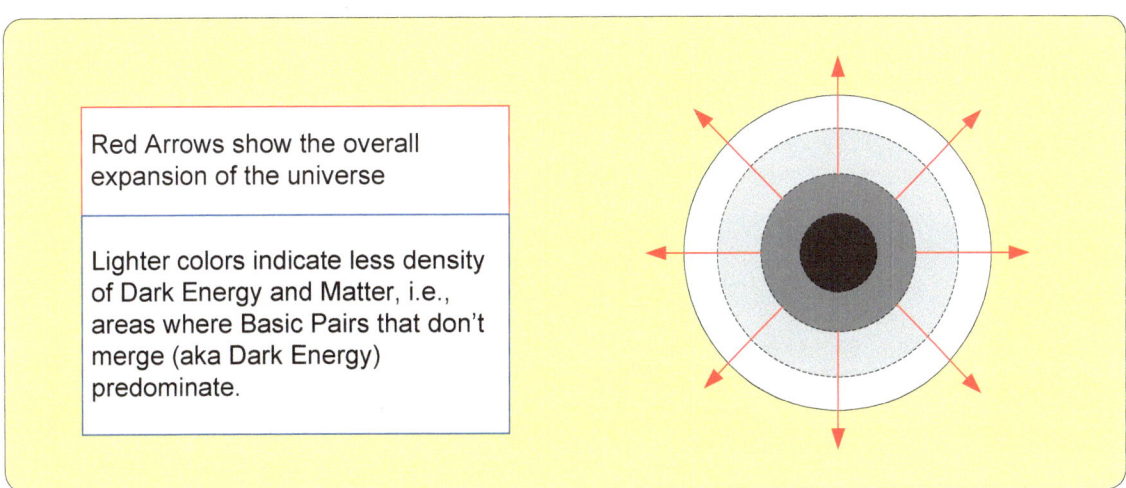

Figure 2-1 – Post-Big Bang Expanding Universe

Note that the center of Figure 2-1 is where the Big Bang occurred. Baryonic Matter, Dark Matter and Dark Energy are initially most dense closer to the initial Singularity. As will be seen below, the outer edges of this diagram represent the areas where Dark Energy is least concentrated.

Universe continues to expand. Baryonic and Dark Matter continue to be less and less dense as they get farther and farther from the Singularity.

Dark Energy (which is consists of Basic Pairs) is less dense near the "leading edge" of the "rising bread loaf" of the post-Big Bang Universe, which continues to expand.

DE has a non-zero vacuum expectation value (VEV) because it consists of BPs, which have energy. The expansive post-Big Bang forces speed up as the non-zero VEV is closer and closer to zero farther and farther from the initial singularity.

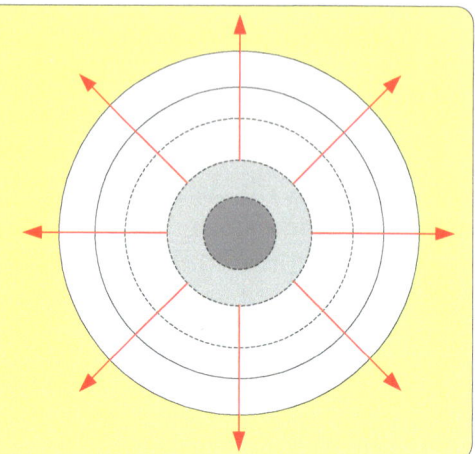

Figure 2-2 – Current Expanding Universe

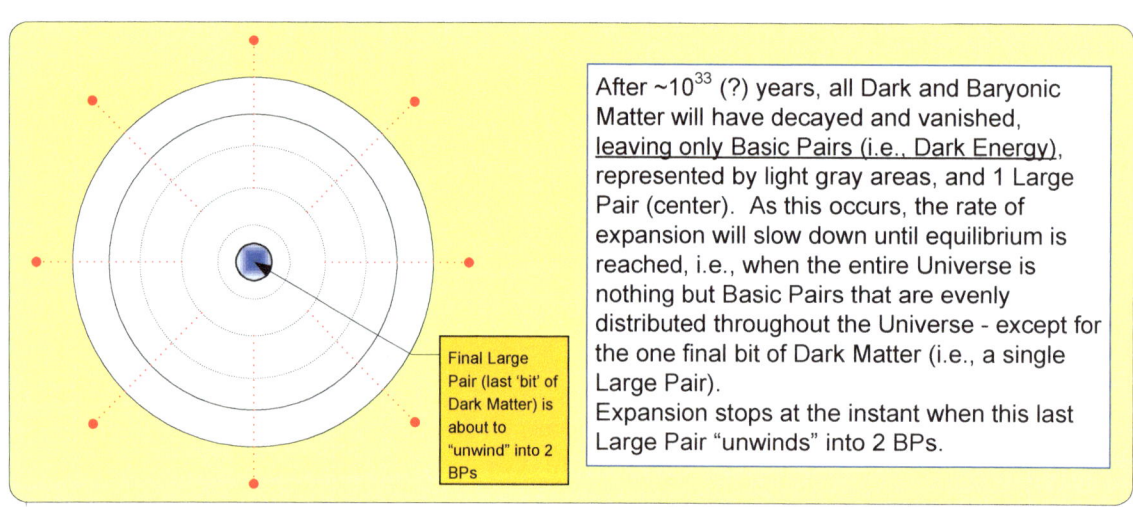

After ~10^{33} (?) years, all Dark and Baryonic Matter will have decayed and vanished, leaving only Basic Pairs (i.e., Dark Energy), represented by light gray areas, and 1 Large Pair (center). As this occurs, the rate of expansion will slow down until equilibrium is reached, i.e., when the entire Universe is nothing but Basic Pairs that are evenly distributed throughout the Universe - except for the one final bit of Dark Matter (i.e., a single Large Pair).

Expansion stops at the instant when this last Large Pair "unwinds" into 2 BPs.

Final Large Pair (last 'bit' of Dark Matter) is about to "unwind" into 2 BPs

Figure 2-3 – Universe's Expansion Stops – The Final Large Pair

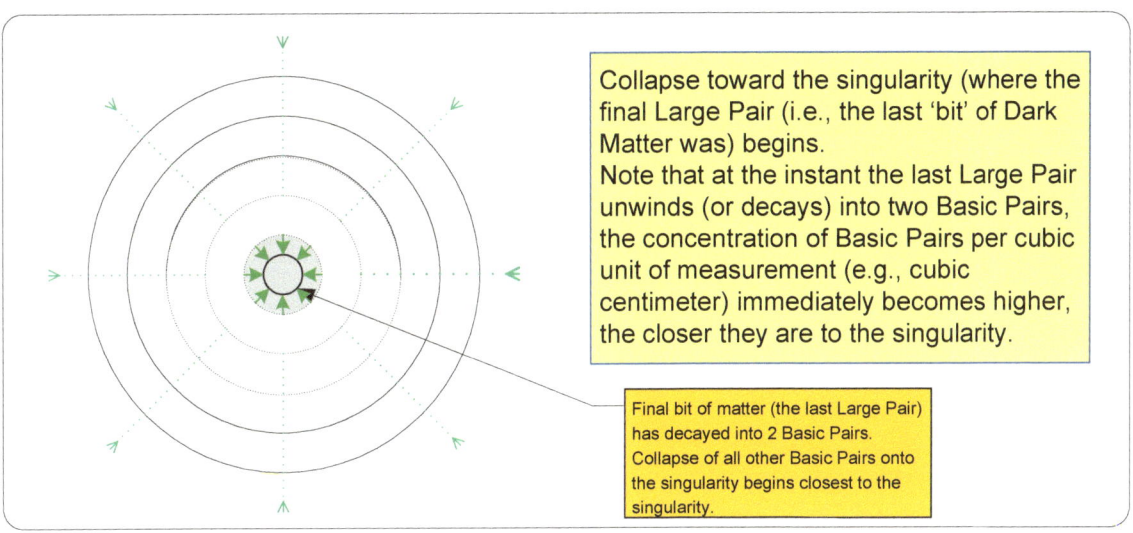

Figure 2-4 – The Collapse Begins

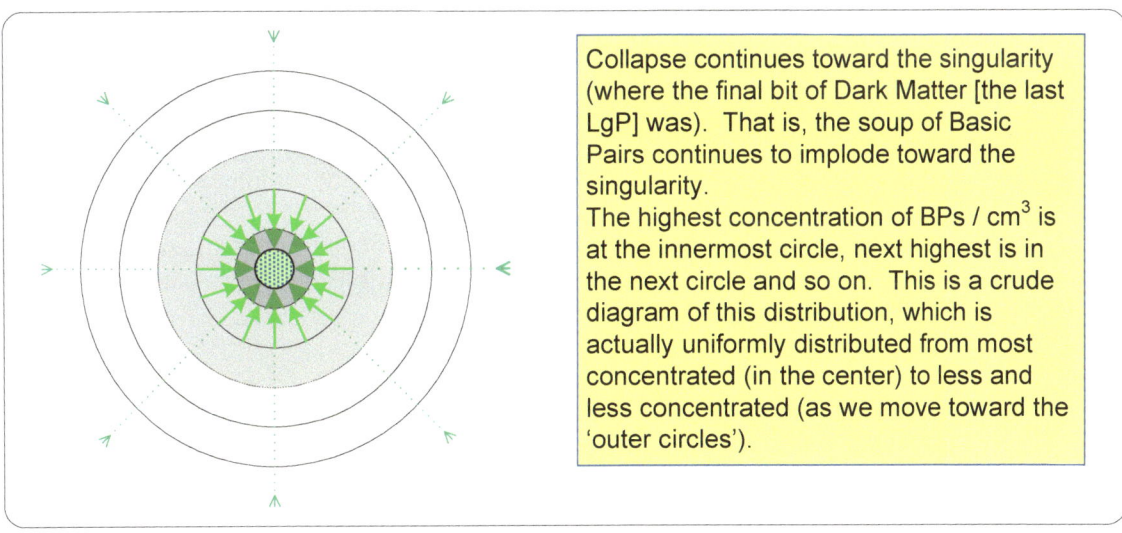

Figure 2-5 – The Collapse Continues

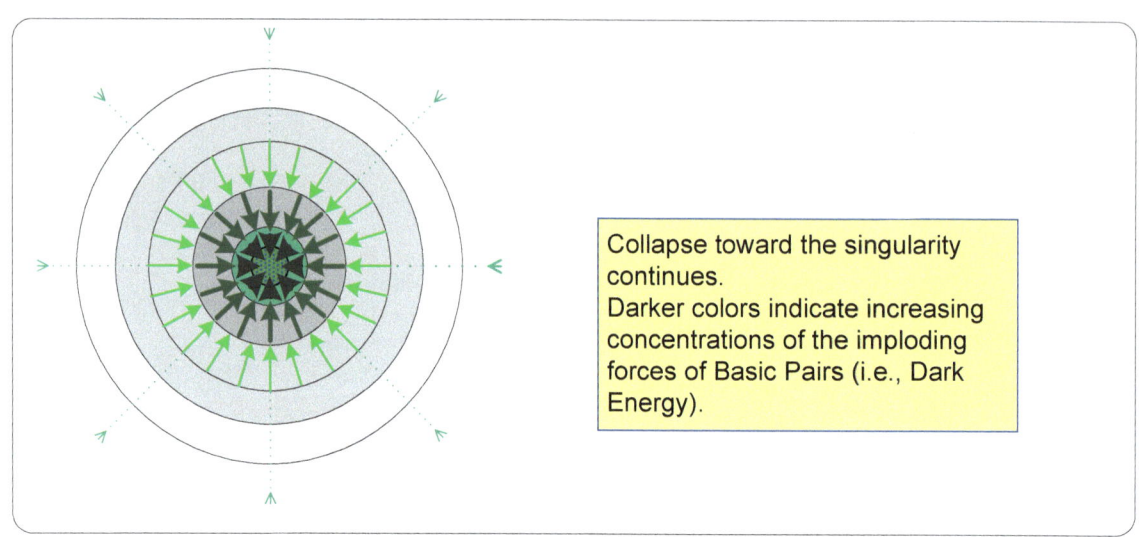

Figure 2-6 – The Collapse Continues Further

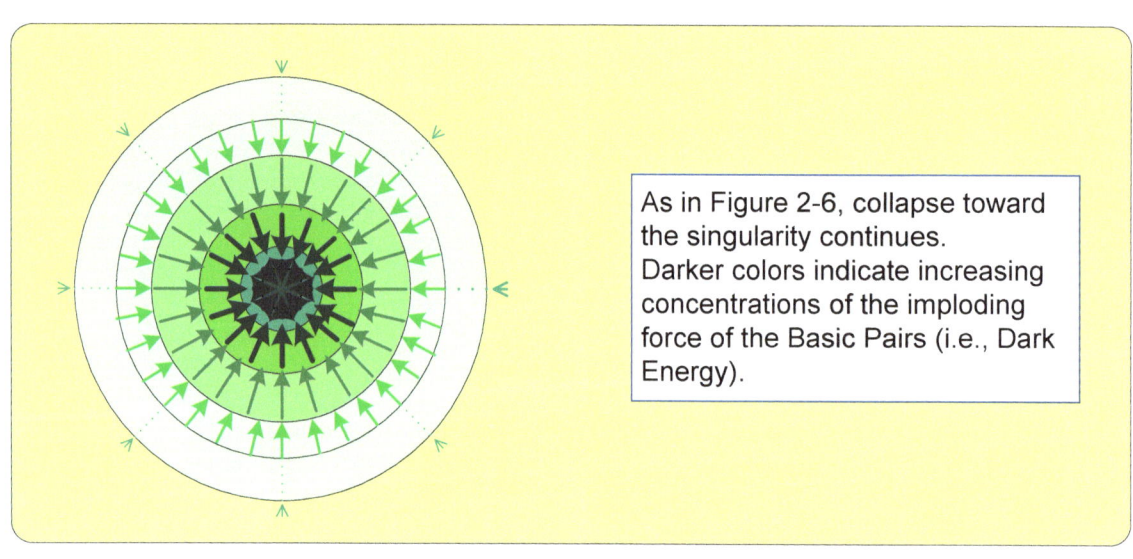

Figure 2-7 – The Collapse Continues On

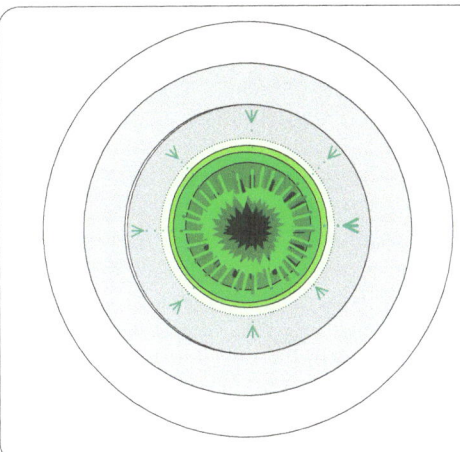

Figure 2-8 – Penultimate Diagram of Collapse

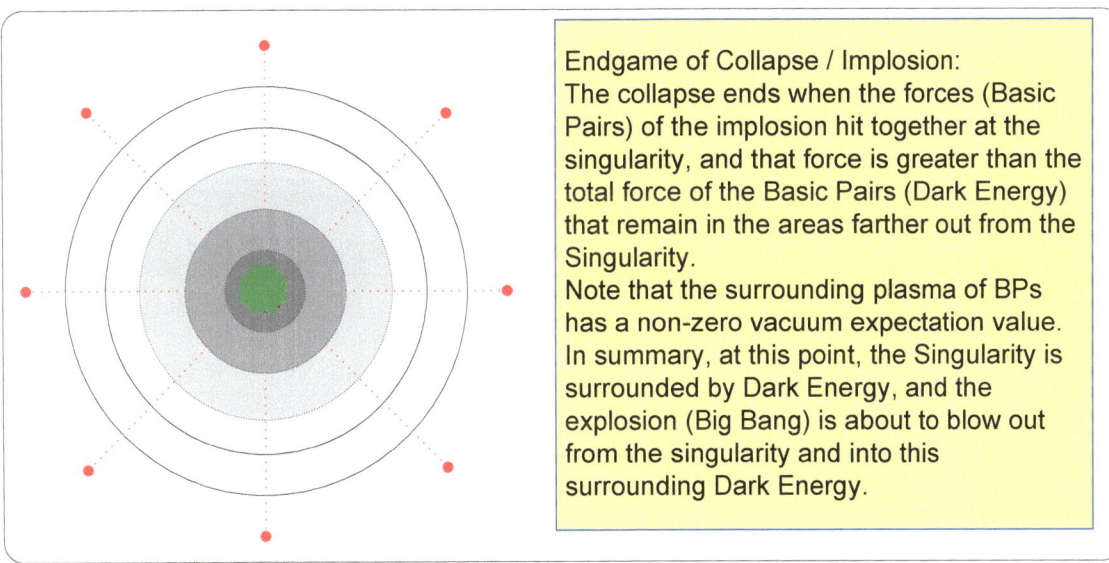

Figure 2-9 – Endgame of Collapse / Implosion

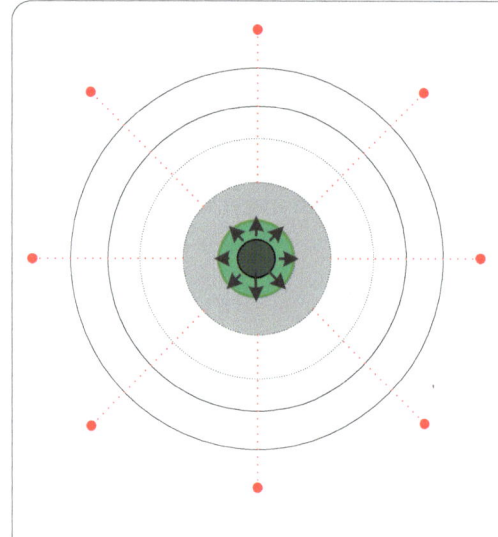

Figure 2-10 – Big Bang Occurs

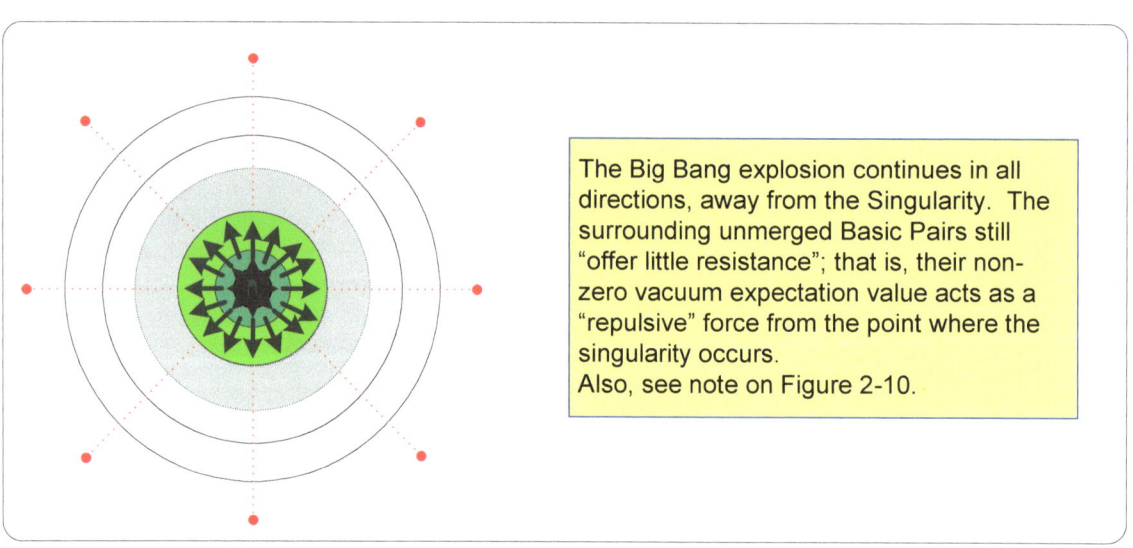

Figure 2-11 – Big Bang Continues

Figure 2-12 below shows a current model of the "Epochs" that begin to roll out after the Big Bang. These Epochs are the same as the sequence started in Figure 2-10 and continuing in 2-11.

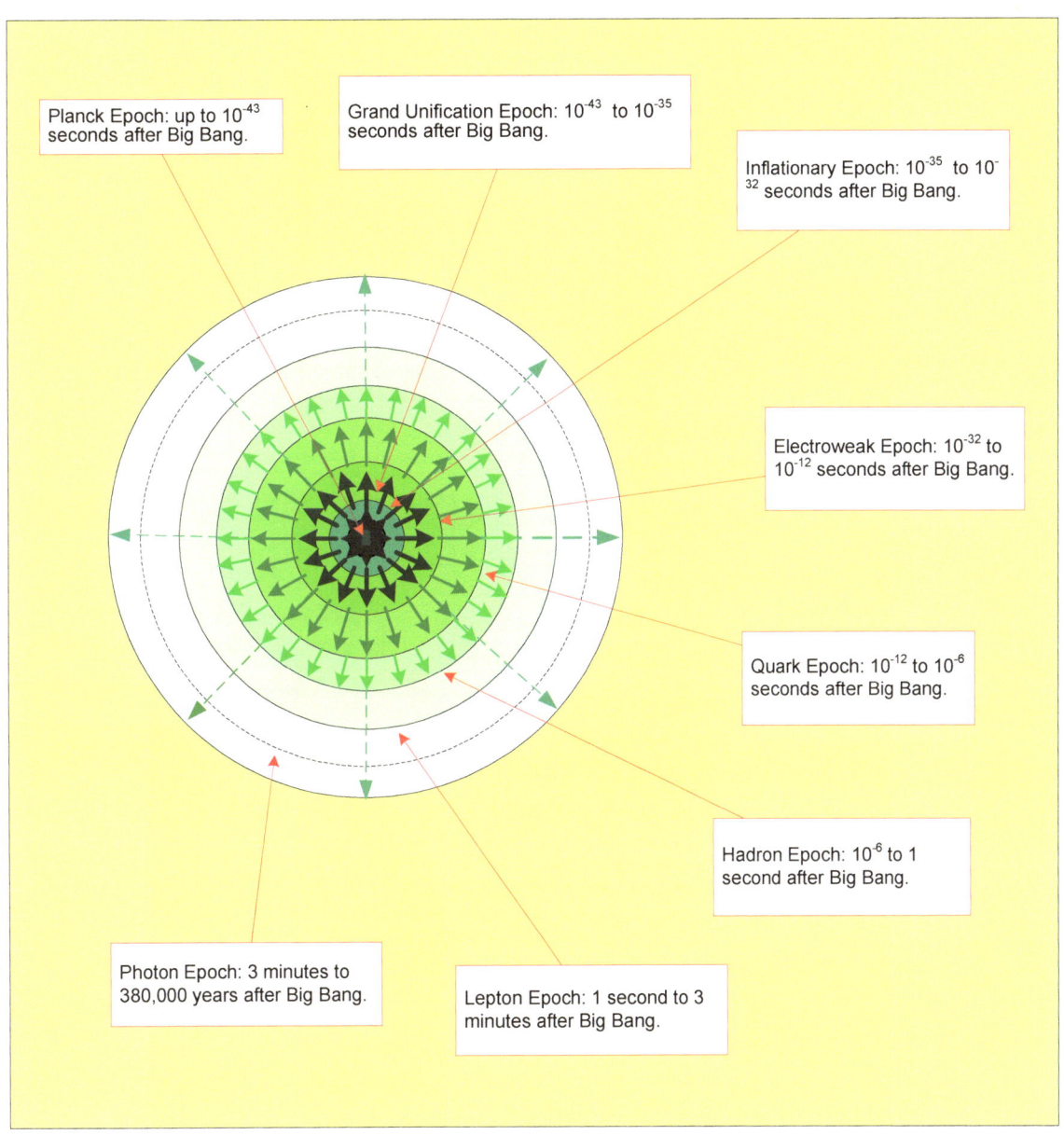

Figure 2-12 – Post-Big Bang Epochs

Figure 2-13 below shows the emergence of the four forces – Gravity, Strong Nuclear Electromagnetic, and Weak Nuclear - during these early phases after the Big Bang. This is not drawn to any scale.

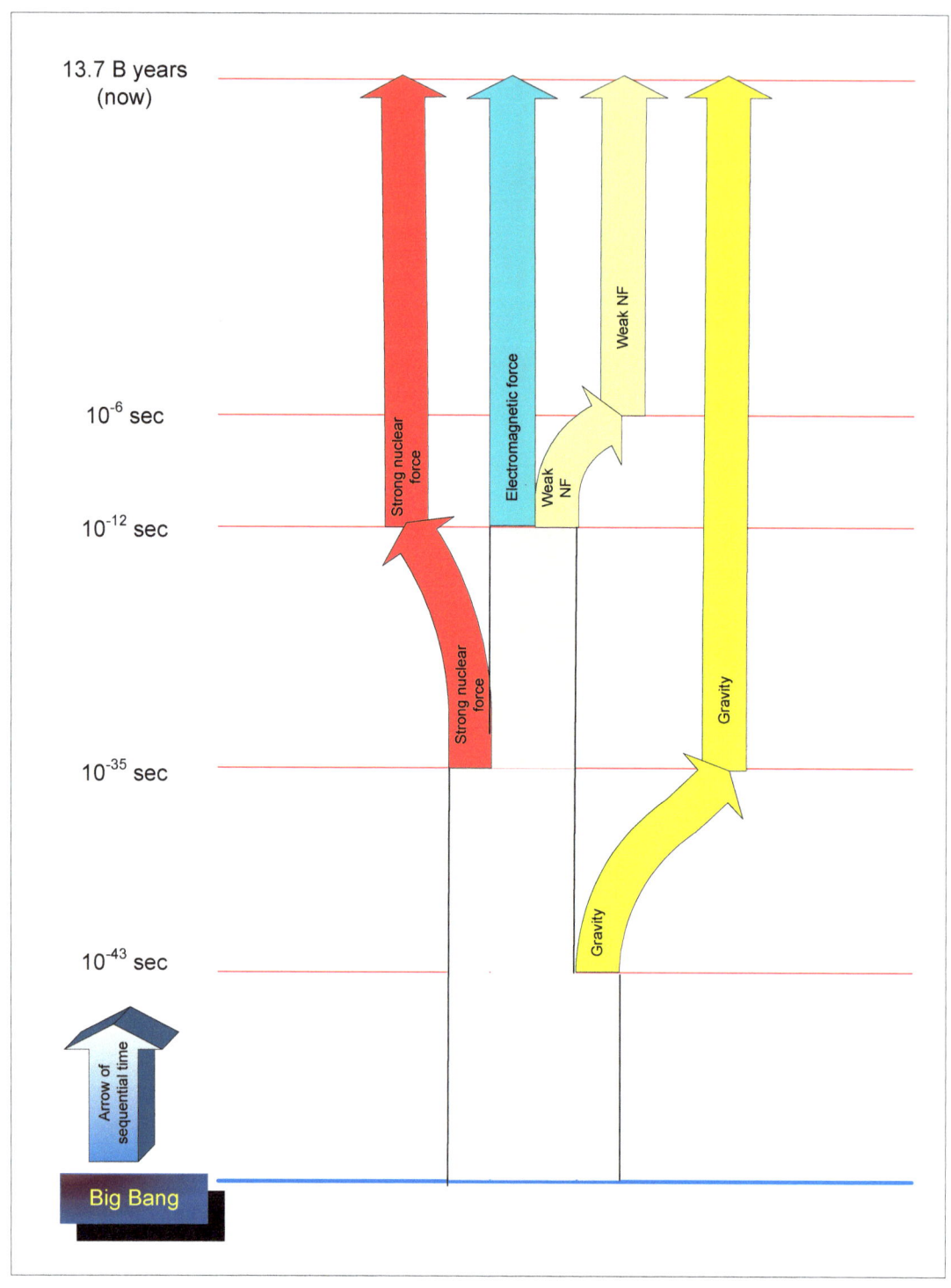

Figure 2-13 – Emergence of the Four Forces

Figure 2-14 below integrates:
- The names and durations of the post-Big Bang Epochs
- Some of the key events that occurred during each Epoch
- The evolution of the Four Fundamental Forces

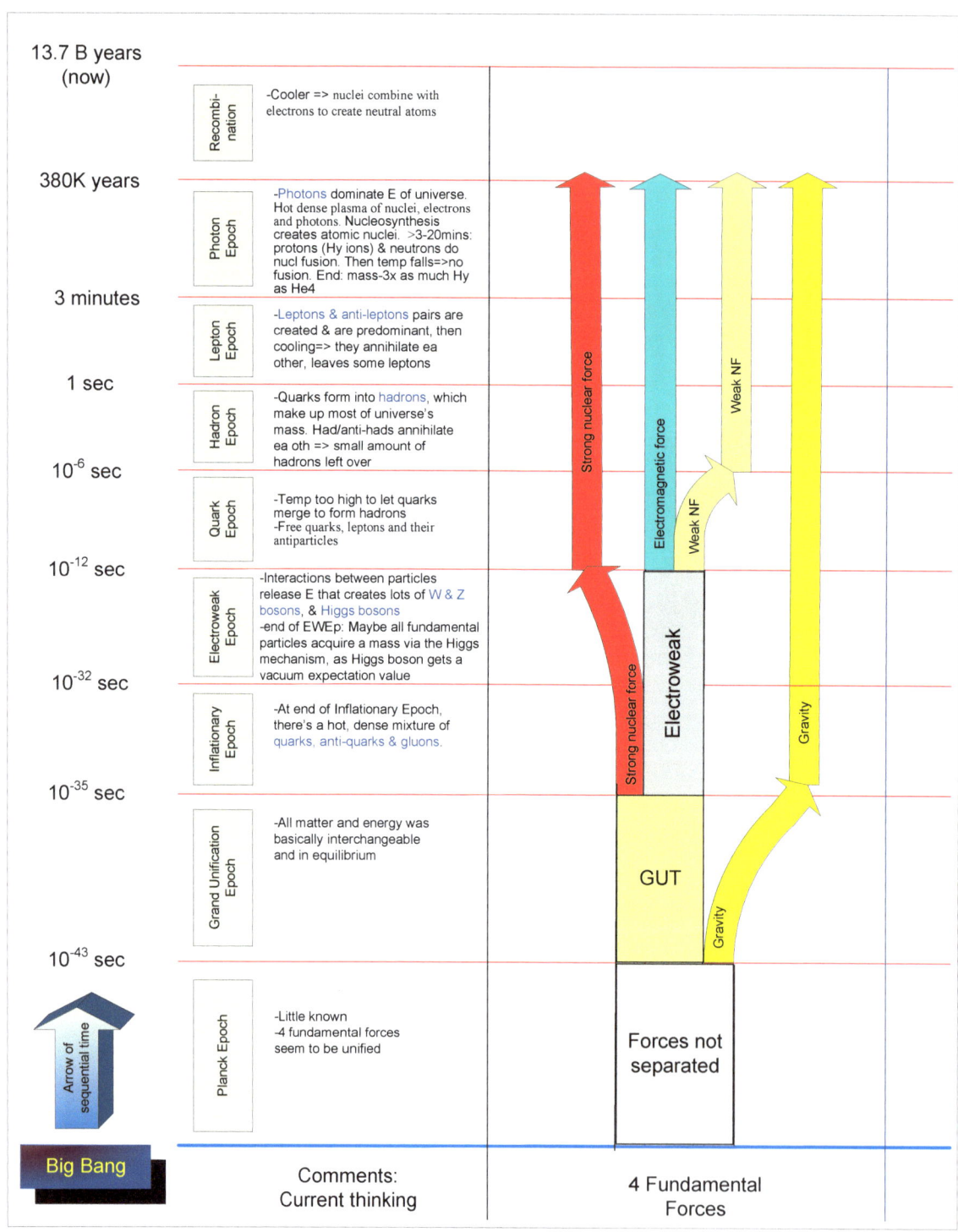

Figure 2-14 – Post-Big Bang Epochs & Events

The question arises: If these epochs and their durations are somewhat understood today, and if they are fairly well represented in Figures 2-12, 2-13 and 2-14, then when do the interactions described in Part 1 of this paper take place, i.e., the interactions of Basic Pairs, Large Pairs and so on?

****

Implications of this Proposal:

Part 1 includes the following hypotheses:
- That Absolute Nil and Non-Nil exist.
- That they exist only as Basic Pairs.
- That Basic Pairs interact in certain ways, as outlined in Part 1.

Parts 1 and 2 together indicate that the Basic Pairs are never destroyed from the instant of the most recent ("first") Big Bang (the one that happened ~13.7 billion years ago), subsequent expansion of the Universe, through the implosion to the singularity, at the point of the second Big Bang, throughout the next expansion of the Universe, and so on (see Figures 2-1 through 2-11 above).

Given the model laid out in Part 1 of this paper, I propose the following:
At the instant when the Big Bang starts, the model described in Part 1 is reset to "0", i.e., to what I describe as the "Initial State" in Part 1. This means that the sequence of Basic Pair interactions begins at the instant of the Big Bang, which is the 'starting place' (or "Initial Situation") described in *Section 3.5.1 - Interactions that Follow the Initial Situation – Phase 1*.

As described in Part 1 of this paper, the Basic Pair interactions and the subsequent interactions of Large Pairs, Very Large Quads, and Very Large Octets lead to the creation and measurable distribution of types of Energy / Matter. My model stops after the creation of XL-12's (i.e., the end of what I call Phase 4):

In short, after Phase 4, the singularity where a Big Bang (either "#1" or "the next one - #2") occurs is surrounded by a "plasma" of billions of Basic Pairs that will no longer interact (per the Part 1 definition of "interact") with each other. I propose, then, that the Phases 1 through 4 interactions (see Part 1) occur in between 1) the instant when a Big Bang occurs and 2) before the end of the Inflationary Epoch, i.e., 10^{-32} second after the Big Bang occurs.

Note that at the end of the Inflationary Epoch, the Universe most likely consists of a hot, dense plasma of quarks, anti-quarks and gluons. I propose that the four phases of the Codex 5 Model occur before this plasma and the quarks, anti-quarks and gluons form.

This means that the Very Large Octets (VLO's) and XL-12's are precursors to quarks and all other elementary particles that ultimately form baryonic matter. It appears that in some way, VLO's and XL-12's interact and are transformed to eventually form quarks and possibly leptons.

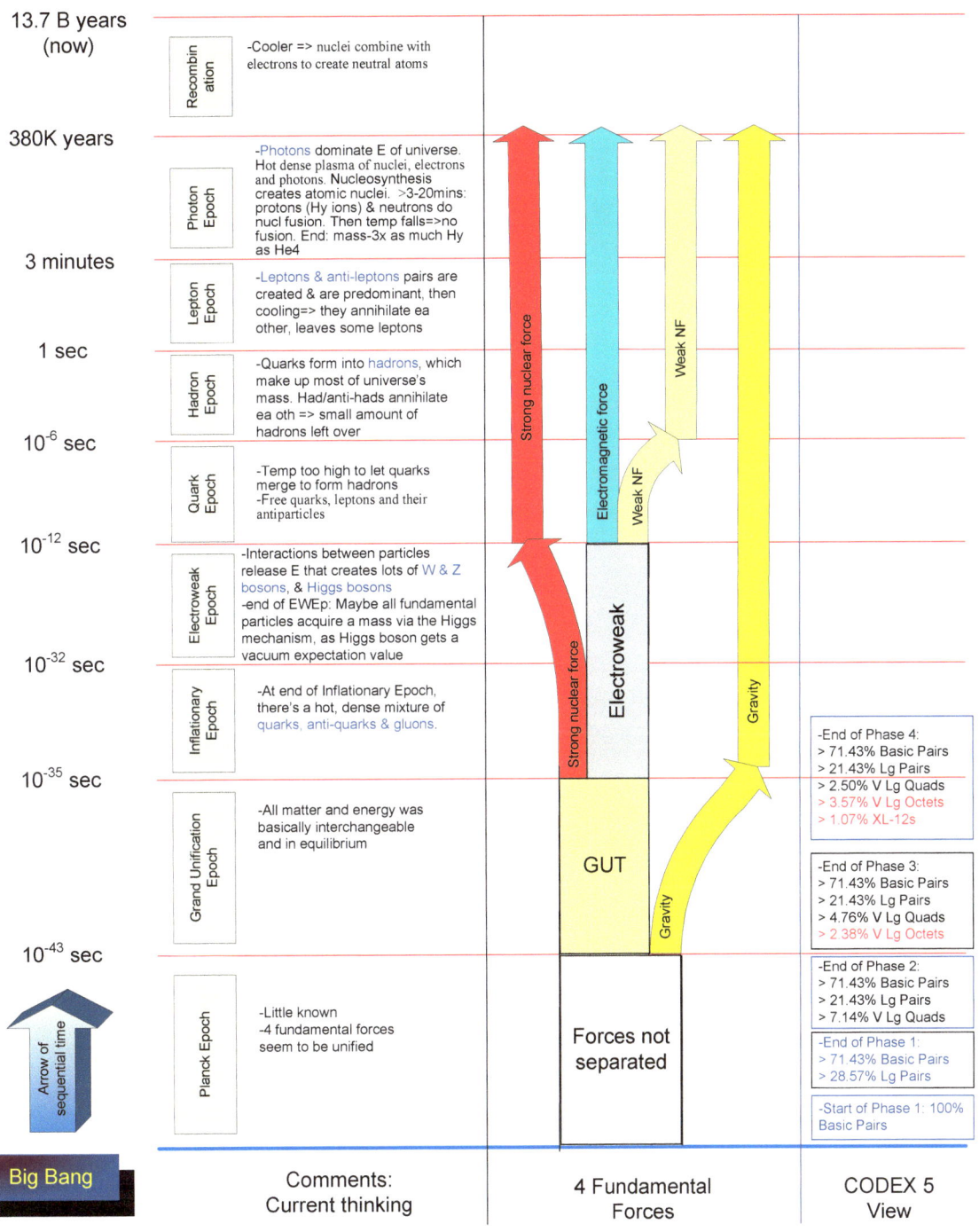

Figure 2-15 – Mapping CODEX 5 Model to Big Bang Epochs & Events

Table 10 (below) presents the Figure 2-15 data in tabular format.

Table 10 – Mapping Codex 5 Model to Big Bang Epochs and Events

	Basic Pairs	Large Pairs	Very Lg Quads	Very Lg Octets	XL-12s	EPOCH	Characteristics of Epoch	Comparison of CODEX 5 with Epoch Events
Start Phase 1	100.00%	0.00%	0.00%	0.00%	0.00%	Planck	4 fundamental forces are or at least seem to be unified	At instant of Big Bang and for <10^{-43} second afterward, only Basic Pairs existed. This matches the state of the 4 forces being unified, and with no differentiation between "types" of particles, matter or energy.
Start Phase 2	71.43%	28.57%	0.00%	0.00%	0.00%	Grand Unification	>All matter and energy was basically interchangeable and in equilibrium >Gravity separates from the other forces	During Codex 5 Phase 1, BPs interact, which leads to creation of some Large Pairs (LgP). As LgPs are what Codex 5 calls "Light" Dark Matter, which Gravity acts on, the emergence of LgPs may be associated with the separation of Gravity from the other forces during the Grand Unification Epoch. LgPs don't correlate to any listed particles
Start Phase 3	71.43%	21.43%	7.14%	0.00%	0.00%	Inflationary	At end of Inflationary Epoch, there's a hot, dense mixture of quarks, anti-quarks & gluons.	During Codex 5 Phase 2, Very Large Quads are formed. VLQs are what Codex 5 calls "Heavy" Dark Matter. In Codex 5, Heavy DM (VLQs) ends up making up 2.5% of all Universe's matter, which may correlate to others' neutrino estimates of 0.1% - 5%
Start Phase 4	71.43%	21.43%	4.76%	2.38%	0.00%			During Codex 5 Phase 3, Very Large Octets are formed
Start Phase 5	71.43%	21.43%	2.50%	3.57%	1.07%			During Codex 5 Phase 4 (the last phase), XL-12s are formed
						Electroweak	>Interactions between particles release E that creates lots of W & Z bosons, & Higgs bosons >End of EWEp: Maybe all fundamental particles acquire a mass via the Higgs mechanism, as Higgs boson gets a vacuum expectation value	
						Quark	>Temp too high to let quarks get confined into hadrons >Free quarks, leptons and their antiparticles	
						Hadron	Quarks form into hadrons, which make up most of universe's mass. Had/anti-hads annihilate ea oth => small amount of hadrons left over	
						Lepton	Leptons & anti-leptons pairs are created & are predominant, then cooling=> they annihilate ea other, leaves some leptons	
						Photon	Photons dominate E of universe. Hot dense plasma of nuclei, electrons and photons. Nucleosynthesis creates atomic nuclei. >3-20mins: protons (Hy ions) & neutrons do nucl fusion. Then temp falls=>no fusion. End mass~3x as much Hy as He4	
						Recombination	Cooler => nuclei combine with electrons to create neutral atoms	

The mechanics of how VLO's and XL-12's may become transformed into quarks and leptons are open to speculation. See pages 57-60 below for a possible explanation of how this transformation occurs. Note also that since VLO's are smaller (in terms of Energy / Mass) than XL-12's, they might be candidates to form electrons, which have a lower mass than up quarks – and likewise, the XL-12's, which are larger than VLO's, could transform into the larger Up Quarks.

Appendix 1 – Codex 5: Implications

The Codex 5 Model addresses the questions listed below...

- 1a) Dark Energy: What does it consist of? Where did it come from?

- 1b) Dark Matter: What does it consist of? Where did it come from?

- 2) How were elementary particles created? The Codex 5 Model describes this process at the lowest existing energy levels.

- 3) Density of Vacuum: For the inflation of the Universe after the Big Bang, a surrounding non-zero vacuum expectation value (VEV) is required. How does this non-zero VEV come about?

- 4) Inflation: Why and how does matter/energy 'expand' from the initial singularity of the Big Bang?

- 5) Cosmological Constant: What is Λ?: $\underline{\Lambda = 0}$, or $\underline{\Lambda > 0}$, or $\underline{\Lambda < 0}$?

- 6) "The Edge of the Universe": The Universe is expanding, so what is on the "outer edge"? Where is it expanding to?

- 7) The Singularity: What caused the Singularity where the Big Bang occurred?

See details on following pages

Pages 52 through 54 summarize the areas listed above.

1A - Dark Energy

Hypothesis: Dark Energy is Basic Pairs
CODEX: Basic Pairs ~ 71.43% of Energy/Matter in universe
WMAP: Dark Energy ~ 72%

The energy/mass of BPs is always in the fabric of space. For detailed sequence, See Codex 5.

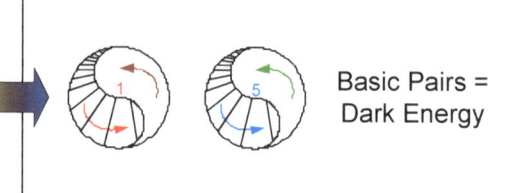

Basic Pairs = Dark Energy

1B - Dark Matter

Hypothesis: Dark Matter is Large Pairs & Very Large Quads
CODEX: LgPs + VLQs ~23.93% of Energy/Matter in universe
WMAP: Dark Matter ~23%
>Large Pairs ~ 21.43% of Energy/Matter ("Light" Dark Matter)
>V Lg Quads ~ 2.5% ("Heavy" Dark Matter)

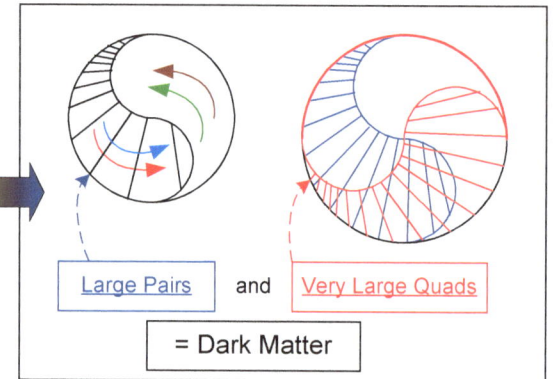

Large Pairs and Very Large Quads

= Dark Matter

	Start Phase 1	Start Phase 2	Start Phase 3	Start Phase 4	Start Phase 5
Basic Pairs	100.00%	71.43%	71.43%	71.43%	71.43%
Large Pairs	0.00%	28.57%	21.43%	21.43%	21.43%
Very Lg Quads	0.00%	0.00%	7.14%	4.76%	2.50%
Very Lg Octets	0.00%	0.00%	0.00%	2.38%	3.57%
XL-12s	0.00%	0.00%	0.00%	0.00%	1.07%

2 - Creation of Elementary Particles

Hypothesis: The plasma/soup of Basic Pairs and Large Pairs described in Codex 5 is similar to, or possibly the equivalent of what's called the Higgs Field.

Like the proposed Higgs Field, the BPs and LgPs don't vanish in a vacuum, i.e., they have a non-zero vacuum expectation value.

Hypothesis: VLOcts and XL12's are "steps along the path" to the development of other particles.

Elementary Particles are eventually formed and acquire their mass after the process outlined in Codex 5. The Codex 5 goes only as far as the formation of the XL-12, but it postulates that the XL-12 and VLO may merge or otherwise evolve into larger particles.

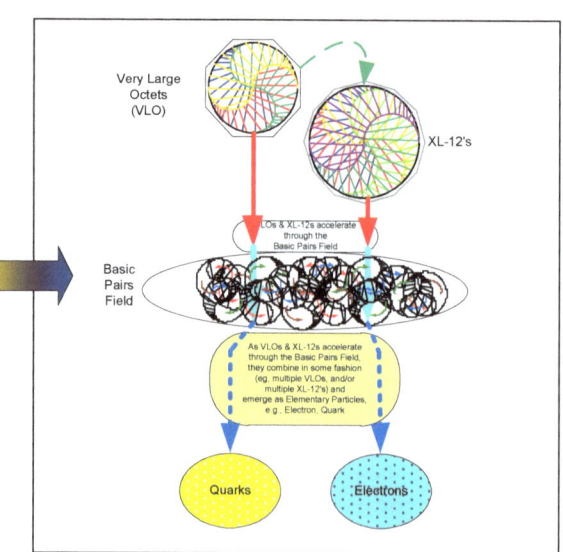

3 - Zero-point Energy / Density of Vacuum

Hypothesis: Existence of Basic Pairs & Large Pairs means that the vacuum expectation value is non-zero.

A positive vacuum energy density means that vacuum's pressure is negative, which leads to an outward expansion from the Big Bang singularity into the surrounding space, which is empty, except for BP and LgPs.

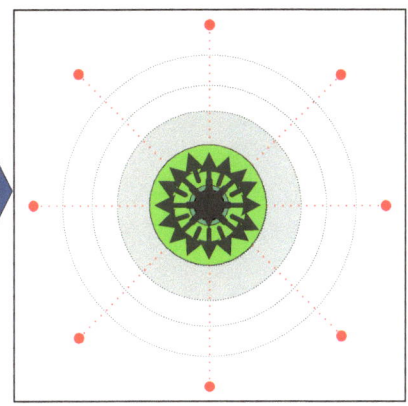

4 - Inflation

Hypothesis: Initial massive inflation was not just a sudden expansion in the size of the universe right after the Big Bang.

The BPs were already surrounding the singularity, but didn't exist relative to anything aside from other BPs. Once the Big Bang happened, LgPs, VLQs, VLOs and XL12s came into existence. So the plasma of pre-Big Bang BPs (around the singularity) suddenly 'existed' relative to these new entities. As events unrolled, new particles were formed (quarks, leptons, etc.), and the pre-Big Bang soup of BPs also 'existed' relative to these particles' existence.

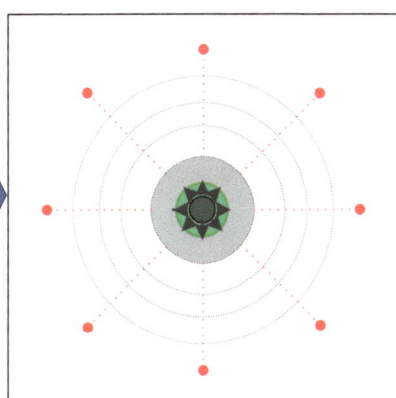

5 - "Cosmological Constant"

Hypothesis: Cosmological Constant changes over time, as follows:

- Since the instant of the Big Bang, the universe is has been expanding at an ever-increasing rate. In this case, $\Lambda > 0$ (aka DeSitter Space).
- At the end of the expansion, just as the last Large Pair of Dark Matter 'unwinds' 'decays' into 2 Basic Pairs (as noted in Codex 5, part 2), $\Lambda = 0$ (Minkowski Space)
- Then, when the implosion toward the singularity point begins, $\Lambda < 0$ (aka Anti-DeSitter Space).

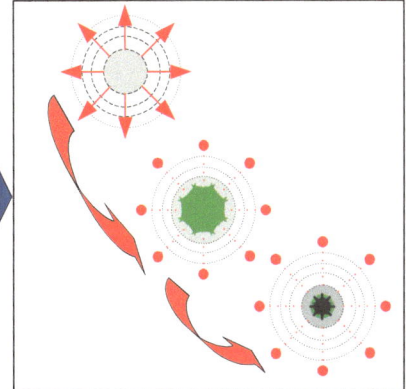

6 - "The Edge of the Expanding Universe"

Hypothesis: After the Big bang the Universe began expanding and is still doing so. Per the Codex 5 Model, Basic Pairs (BPs) surrounded the Singularity at the time of the Big Bang. These BPs have been 'riding the inflationary wave' ever since the Big Bang.

Per the Codex 5, if Absolute Nothingness (called "Ab Nil" in Codex 5) exists, then Non-Nil must also exist. So, if Ab Nil exists beyond the outer edge of the expanding Universe, then when a Basic Pair reaches it, that Ab Nil is subsumed into the BP, per the structure mentioned in Codex 5. The BP retains its same integrity and structure, and proceeds to expand from the initial force of the Big Bang, combined with the complete lack of resistance of the Ab Nil beyond the leading edge of the expansion.

Correlation:
This 'lack of resistance' is equivalent to the force seen as "repulsive" from the Big Bang Singularity, and as "attractive" from the point of view of all that surrounds the Singularity.

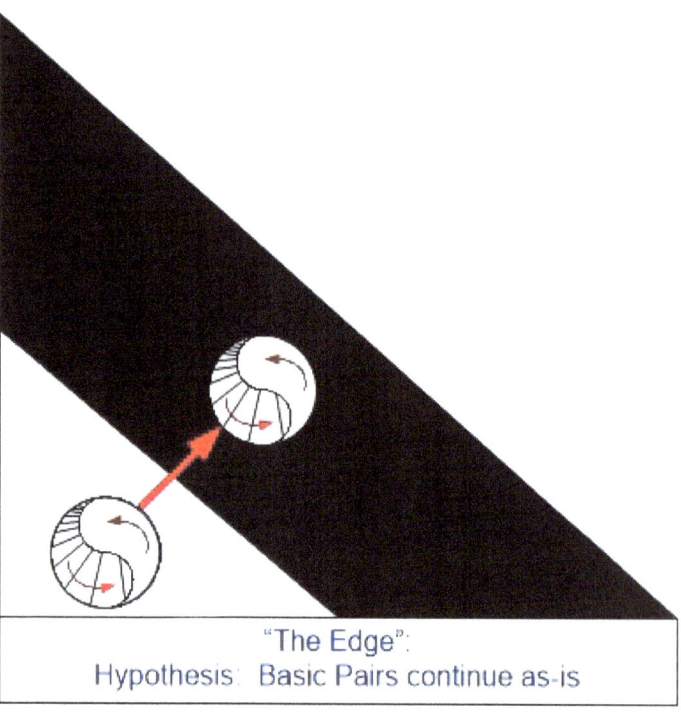

"The Edge":
Hypothesis: Basic Pairs continue as-is

7 – Where the Singularity Came From

Hypothesis: The Singularity where the Big Bang occurs is at the spot where the Last Large Pair 'unwinds" into 2 Basic Pairs.

Correlation:
"...the farther back the fluctuation happened, the lower the entropy it would have had to attain (entropy starts to rise after any dip to low entropy...a small fluctuation early on—a modest jump to the favorable conditions, within a tiny nugget of space—inevitably yields the huge and ordered universe we are aware of."
  B. Green, The Fabric of the Cosmos.

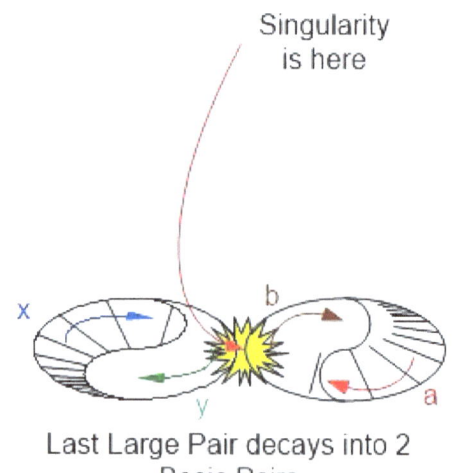

Last Large Pair decays into 2 Basic Pairs

Pages 55 through 65 provide details about the areas summarized above.

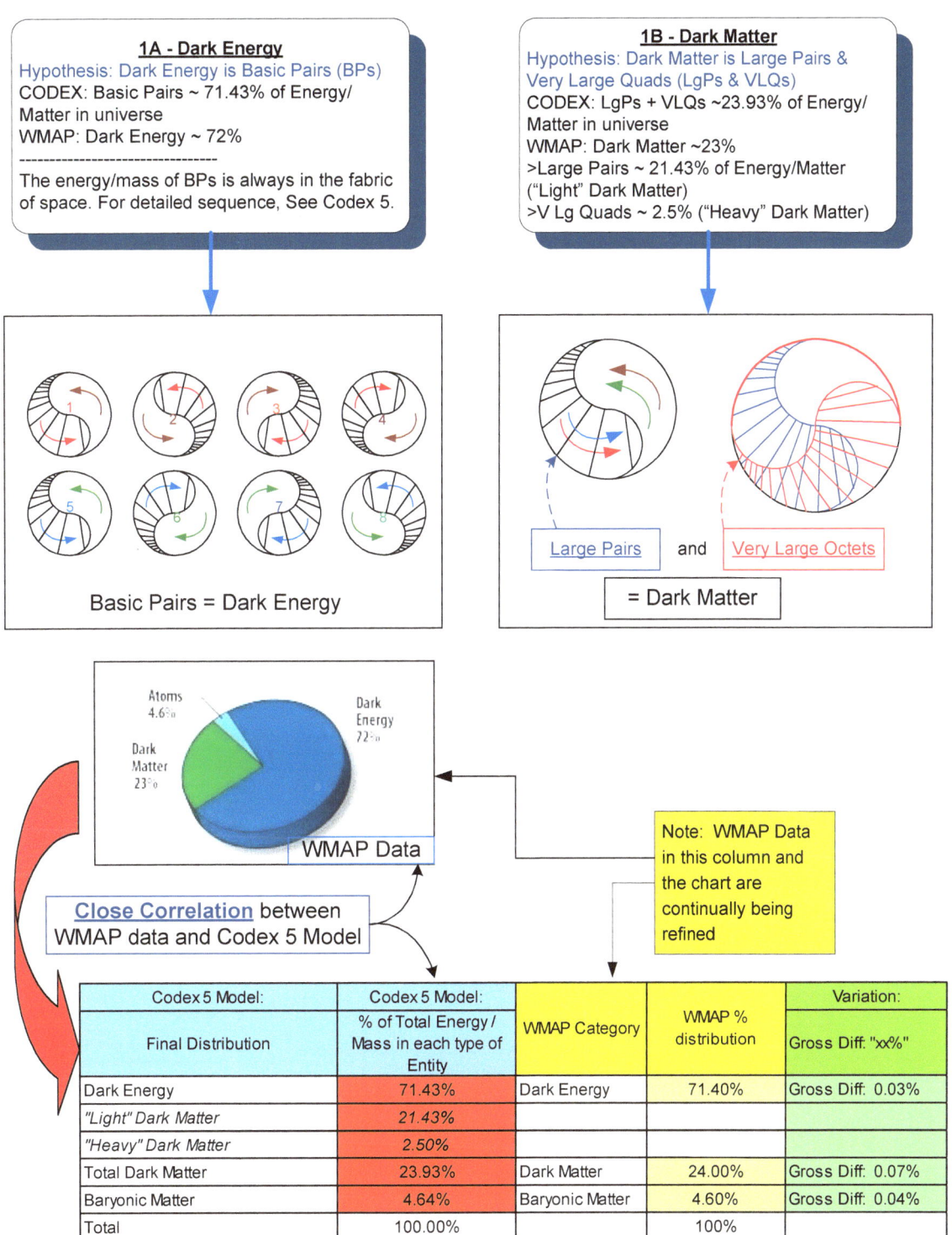

1A - Dark Energy
Hypothesis: Dark Energy is Basic Pairs (BPs)
CODEX: Basic Pairs ~ 71.43% of Energy/Matter in universe
WMAP: Dark Energy ~ 72%

The energy/mass of BPs is always in the fabric of space. For detailed sequence, See Codex 5.

1B - Dark Matter
Hypothesis: Dark Matter is Large Pairs & Very Large Quads (LgPs & VLQs)
CODEX: LgPs + VLQs ~23.93% of Energy/Matter in universe
WMAP: Dark Matter ~23%
>Large Pairs ~ 21.43% of Energy/Matter ("Light" Dark Matter)
>V Lg Quads ~ 2.5% ("Heavy" Dark Matter)

Basic Pairs = Dark Energy

Large Pairs and Very Large Octets = Dark Matter

WMAP Data

Close Correlation between WMAP data and Codex 5 Model

Note: WMAP Data in this column and the chart are continually being refined

Codex 5 Model: Final Distribution	Codex 5 Model: % of Total Energy / Mass in each type of Entity	WMAP Category	WMAP % distribution	Variation: Gross Diff. "xx%"
Dark Energy	71.43%	Dark Energy	71.40%	Gross Diff: 0.03%
"Light" Dark Matter	21.43%			
"Heavy" Dark Matter	2.50%			
Total Dark Matter	23.93%	Dark Matter	24.00%	Gross Diff: 0.07%
Baryonic Matter	4.64%	Baryonic Matter	4.60%	Gross Diff: 0.04%
Total	100.00%		100%	

Note: WMAP figures come from: http://map.gsfc.nasa.gov/universe/uni_matter.html (Nov 2013)

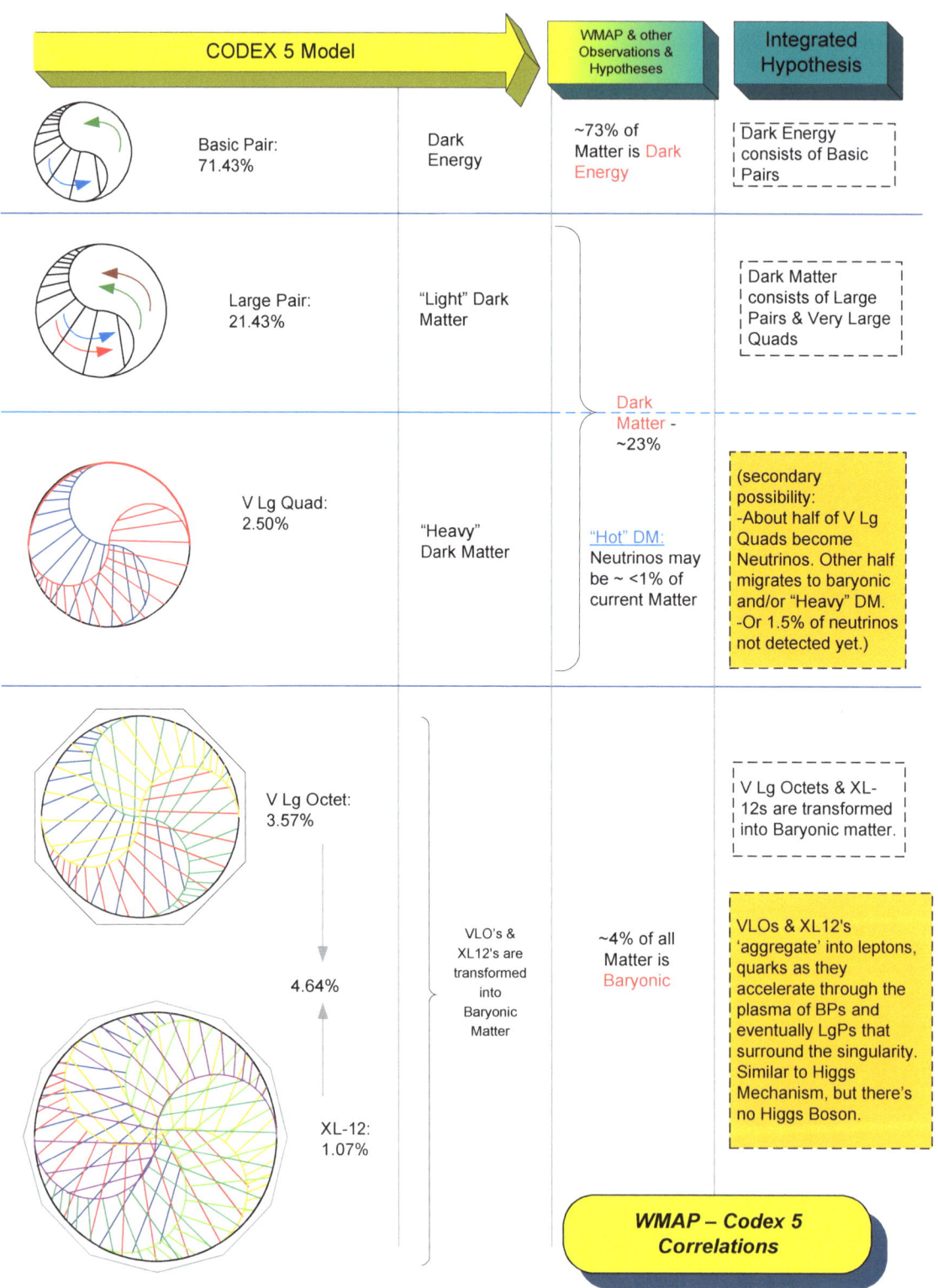

2 - Creation of Elementary Particles

Hypothesis: The plasma/soup of Basic Pairs and Large Pairs described in Codex 5 is similar to, or possibly the equivalent of what's called the Higgs Field.

Like the proposed Higgs Field, the BPs and LgPs don't vanish in a vacuum, i.e., they have a non-zero vacuum expectation value.

Hypothesis: VLOcts and XL12's are "steps along the path" to the development of other particles.

Elementary Particles are eventually formed and acquire their mass after the process outlined in Codex 5. The Codex 5 goes only as far as the formation of the XL-12, but it postulates that the XL-12 and VLO may merge or otherwise evolve into larger particles

Correlations between Codex 5 and Current Theory:

Current Theory: http://www.bun.kyoto-u.ac.jp/~suchii/Leib-Clk/higgs.html:

If that particle *changes its velocity* of movement, that is, if it *accelerates*, then the Higgs field is supposed to exert a certain amount of resistance or drag, and that is the origin of *inertial mass*. In a slightly more precise terminology, inertial mass is generated by *interactions* between a particle and the (nonzero) Higgs field. In a nutshell, this is the origin of inertial mass...Moreover, the degree of resistance (drag) of the Higgs field is different depending on the kinds of fundamental particles, and this generates the difference between the mass of electron and that of a quark.

Codex 5:
See Notes to the left.
See diagrams on this page and the next 2 pages.

See Diagrams on next 2 pages →

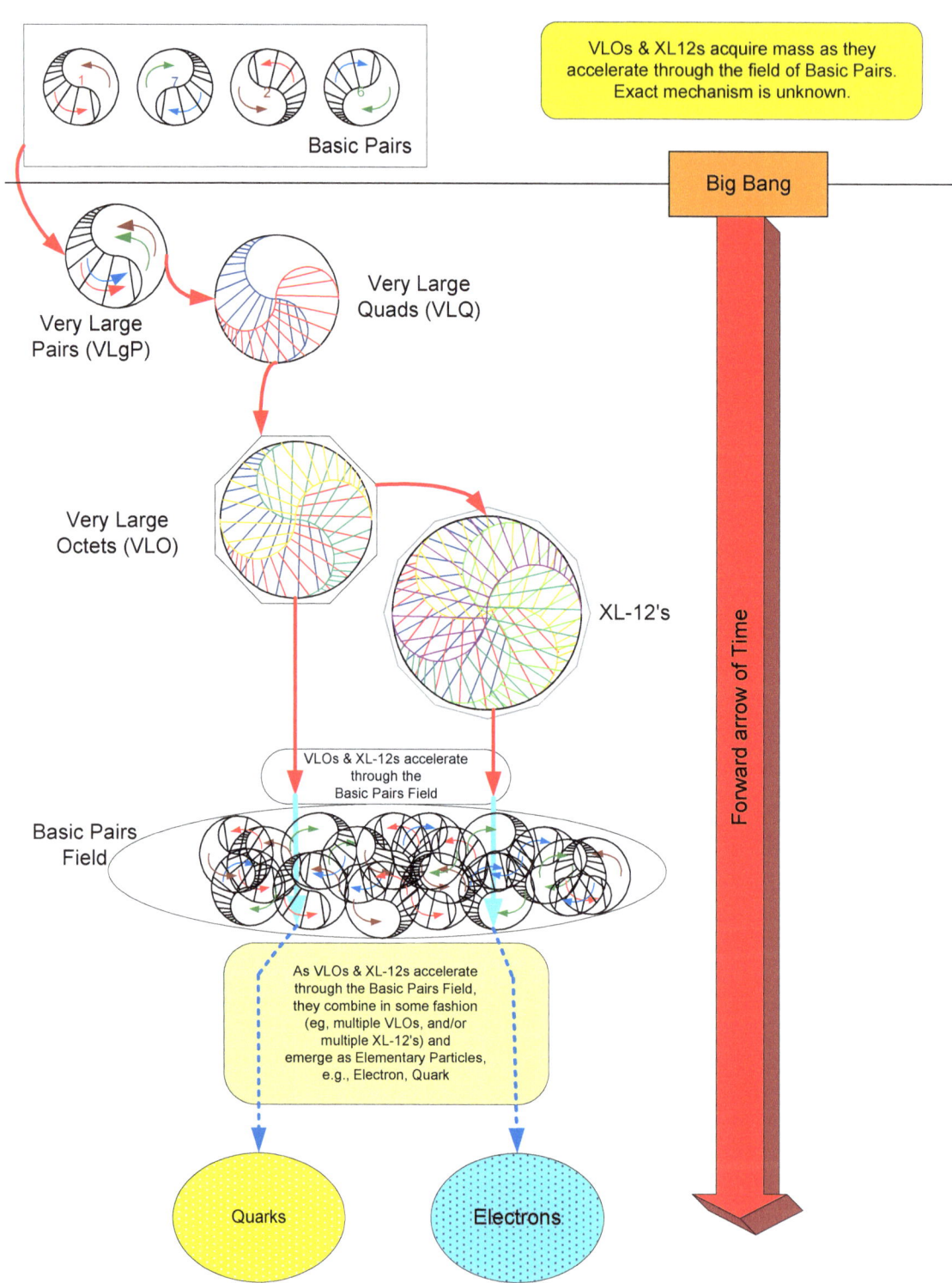

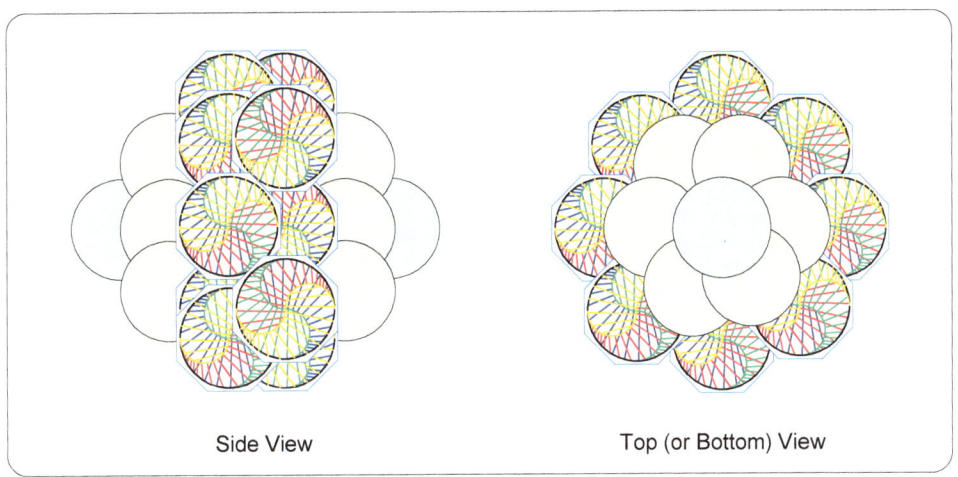

Sample #1 "accretion" of
Very Large Octets (VLO's)

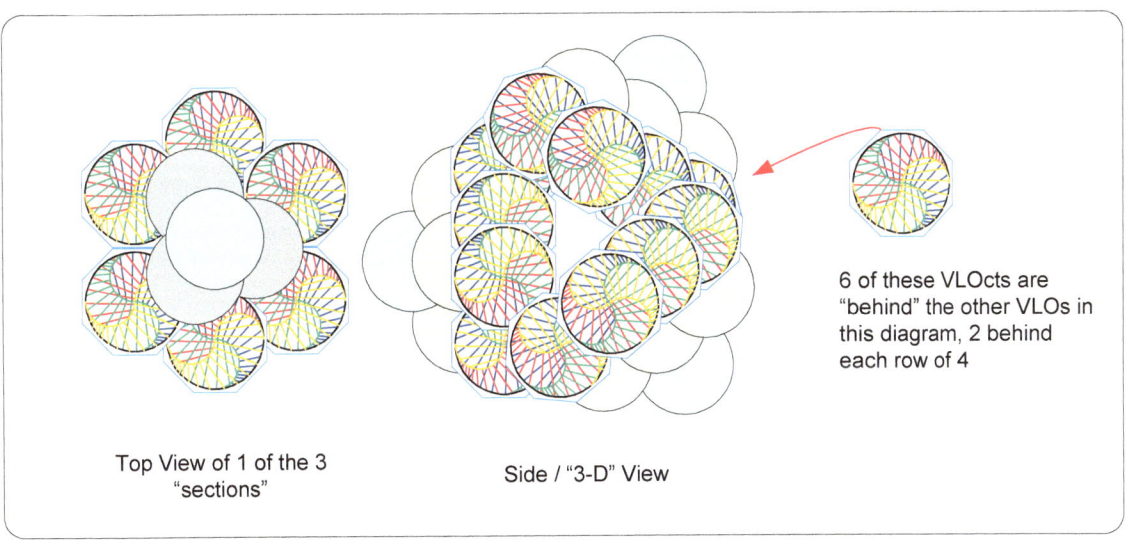

Sample #2 "accretion" of
Very Large Octets (VLO's)

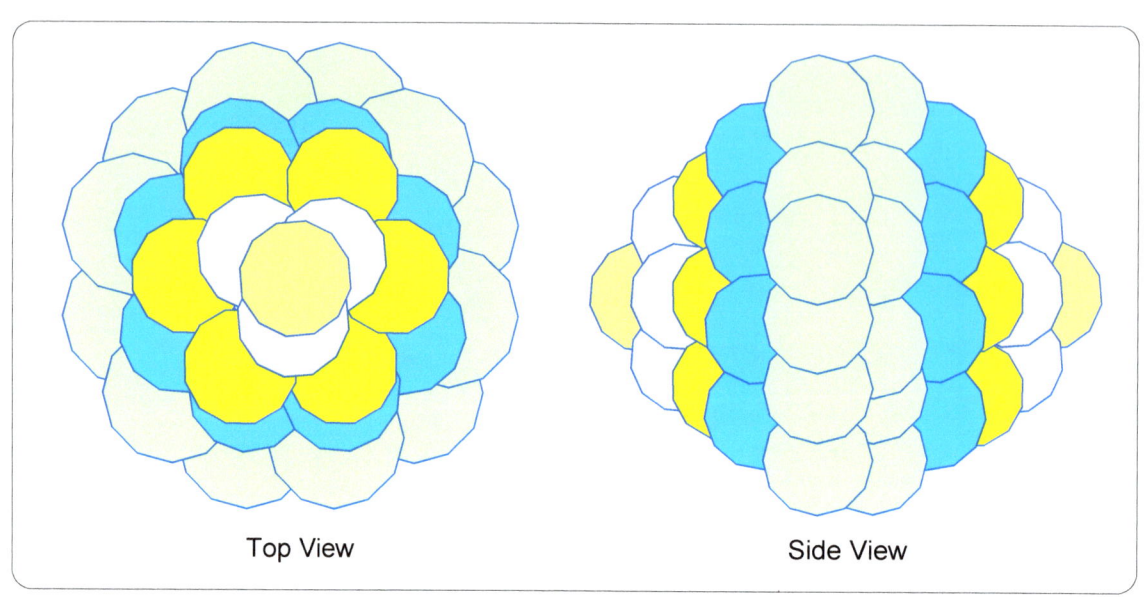

Sample #3 "accretion" of XL-12's

3 - Zero-point Energy / Density of Vacuum

Hypothesis: Existence of Basic Pairs & Large Pairs means that the vacuum expectation value is non-zero.

A positive vacuum energy density means that vacuum's pressure is negative, which leads to an outward expansion from the Big Bang singularity into the surrounding space, which is empty, except for BP and LgPs.

Correlations:
Current Theory: http://math.ucr.edu/home/baez/vacuum.html:
1. We can *measure* the energy density of the vacuum through astronomical observations that determine the curvature of spacetime. All the measurements that have been done agree that the energy density is **VERY CLOSE TO ZERO**. In terms of mass density, its absolute value is less than 10^{-26} kilograms per cubic meter. In terms of energy density, this is about 10^{-9} joules per cubic meter.

... recent measurements by the Wilkinson Microwave Anisotropy Probe and many other experiments seem to be converging on a *positive* cosmological constant, equal to about 6×10^{-27} kilograms per cubic meter. This corresponds to a positive energy density of about 9×10^{-10} joules per cubic meter.

Codex 5:
As noted in the box to the left, the existence of BPs and LgPs means that the energy density of space is >0. Also see Figures below.

Universe continues to expand. Baryonic and Dark Matter continue to be less and less dense as they get farther and farther from the Singularity.

Dark Energy (which is consists of Basic Pairs) is less dense near the "leading edge" of the "rising bread loaf" of the post-Big Bang Universe, which continues to expand.

DE has a non-zero vacuum expectation value (VEV) because it consists of BPs, which have energy. The expansive post-Big Bang forces speed up as the non-zero VEV is closer and closer to zero farther and farther from the initial singularity.

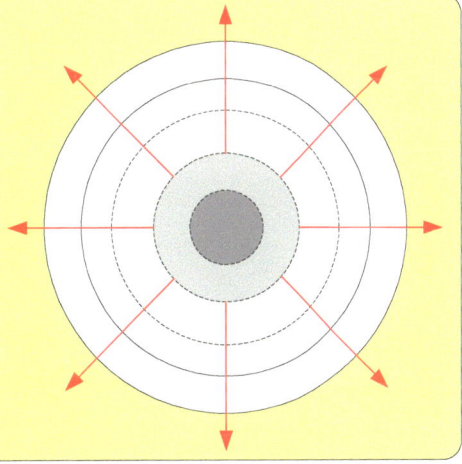

Codex 5, Figure 2-2

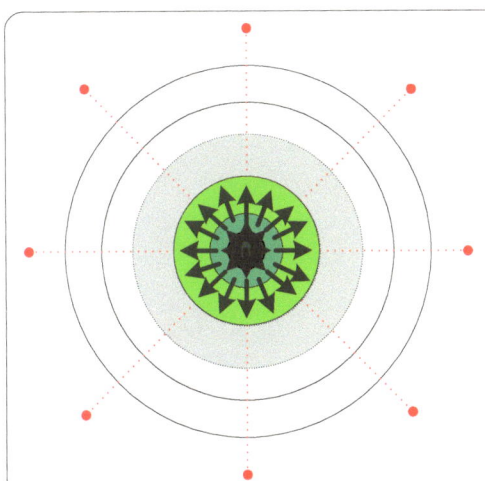

The Big Bang explosion continues in all directions, away from the Singularity. The surrounding unmerged Basic Pairs still "offer little resistance"; that is, their non-zero vacuum expectation value acts as a "repulsive" force from the point where the singularity occurs.

Also, see note on Figure 2-10.

Codex 5, Figure 2-11

4 - Inflation

Hypothesis: Initial massive inflation was not just a sudden expansion in the size of the universe right after the Big Bang.

The BPs were already surrounding the singularity, but didn't exist relative to anything aside from other BPs. Once the Big Bang happened, LgPs, VLQs, VLOs and XL12s came into existence. So the plasma of pre-Big Bang BPs (around the singularity) suddenly 'existed' relative to these new entities. As events unrolled, new particles were formed (quarks, leptons, etc.), and the pre-Big Bang soup of BPs also 'existed' relative to these particles' existence.

Correlations
Current Theory: Inflationary Epoch, 10^{-32} second after Big Bang: Universe expanded by a factor of $\sim 10^{26}$. (ref WMAP data)
Codex 5: See box to left and figures below.

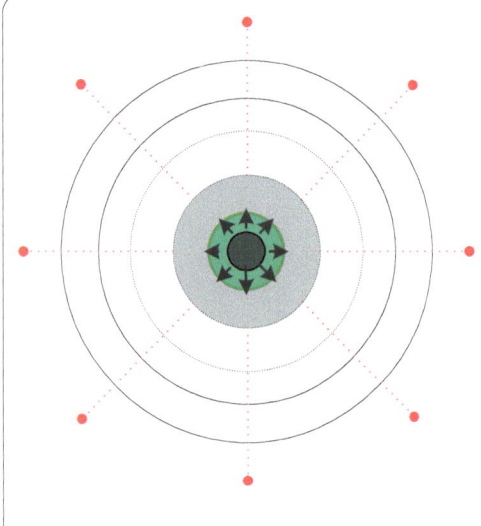

The Big Bang occurs: When enough Basic Pairs' forces implode to and reach a single point (the Singularity), the collision leads to a reactive explosion.

The Singularity is surrounded on all sides by unmerged Basic Pairs (Dark Energy). The total of the Energy of these Basic Pairs is not sufficient to resist the forces of the explosion, and they offer little resistance to the expansion of forces that result from the Big Bang. Note that the Basic Pairs are less and less concentrated toward the outer edges (i.e., farthest from the Singularity), so over time, they – as Dark Energy – will offer less and less resistance to the ongoing expanding forces of what, over time, will become Dark and Baryonic Matter.

Codex 5, Figure 2-10

5 - "Cosmological Constant"

Hypothesis: Cosmological Constant changes over time, as follows:

- Since the instant of the Big Bang, the universe is has been expanding at an ever-increasing rate. In this case, $\Lambda > 0$ (aka DeSitter Space).
- At the end of the expansion, just as the last Large Pair of Dark Matter 'unwinds' 'decays' into 2 Basic Pairs (as noted in Codex 5, part 2), $\Lambda = 0$ (Minkowski Space)
- Then, when the implosion toward the singularity point begins, $\Lambda < 0$ (aka Anti-DeSitter Space).

Correlations:
Current Theory: http://www.astro.ucla.edu/~wright/glossary.html#VED:
vacuum energy density: "Quantum theory requires empty space to be filled with particles and anti-particles being continually created and annihilated. This could lead to a net density of the vacuum, which if present, would behave like a cosmological constant."

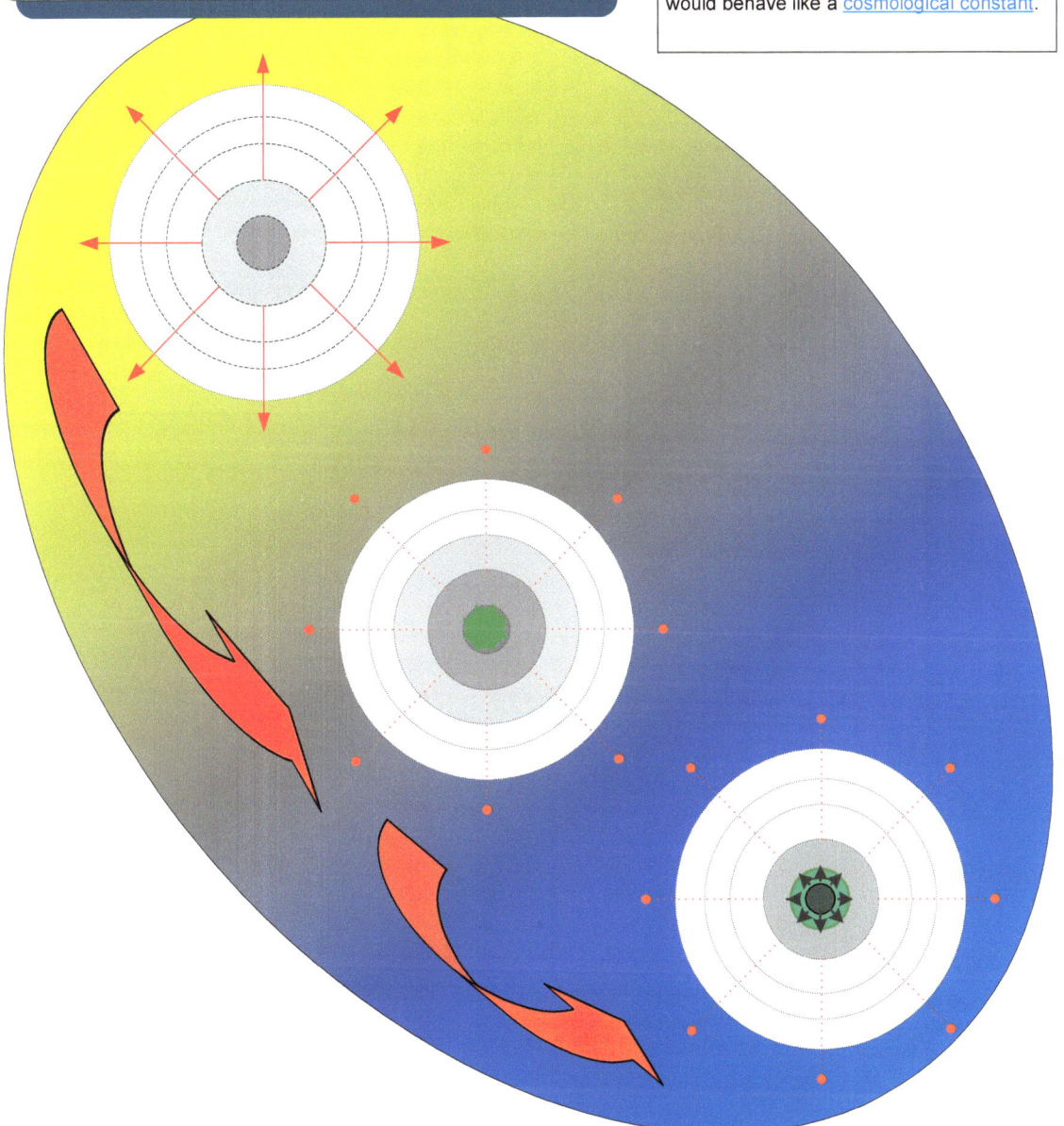

6 - "The Edge of the Expanding Universe"

Hypothesis: After the Big Bang the Universe began expanding and is still doing so. Per the Codex 5 Model, Basic Pairs (BPs) surrounded the Singularity at the time of the Big Bang. These BPs have been 'riding the inflationary wave' ever since the Big Bang. Per the Codex 5, if Absolute Nothingness (called "Ab Nil" in Codex 5) exists, then Non-Nil must also exist. So, if Ab Nil exists beyond the outer edge of the expanding Universe, then when a Basic Pair reaches it, that Ab Nil is subsumed into the BP, per the structure mentioned in Codex 5. The BP retains its same integrity and structure, and proceeds to expand from the initial force of the Big Bang, combined with the complete lack of resistance of the Ab Nil beyond the leading edge of the expansion.

<u>Correlation:</u>
This 'lack of resistance' is equivalent to the force seen as "repulsive" from the Big Bang Singularity, and as "attractive" from the point of view of all that surrounds the Singularity.

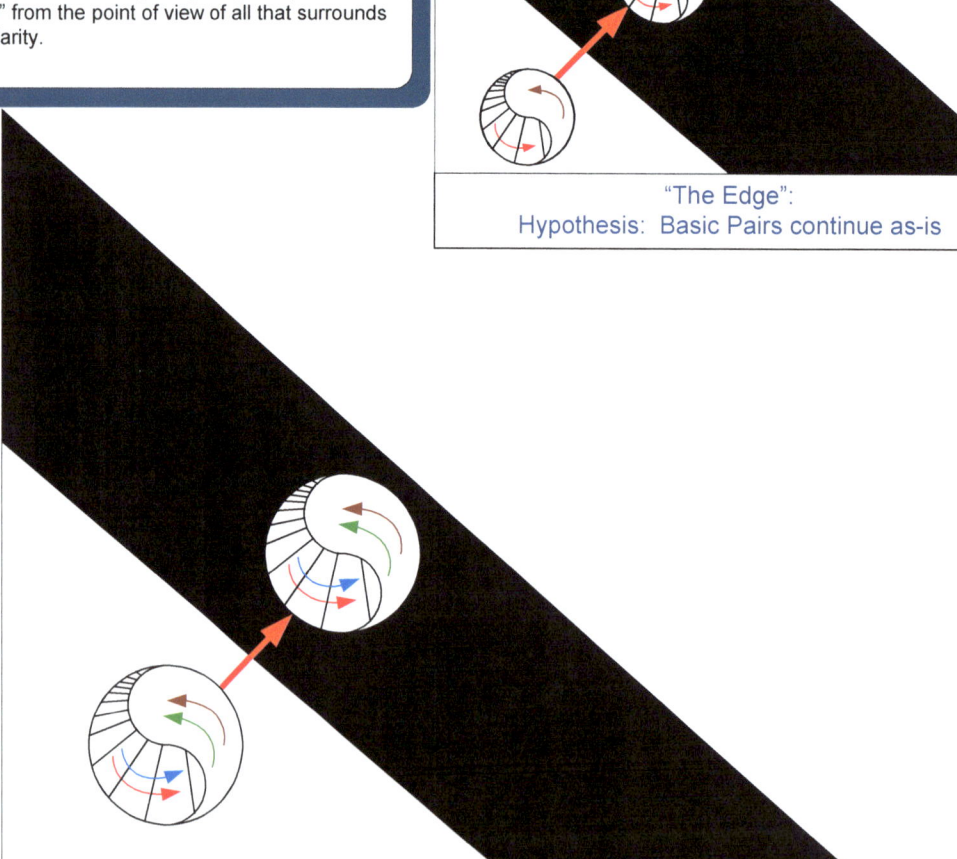

"The Edge":
Hypothesis: Basic Pairs continue as-is

"The Edge":
Hypothesis: Each Large Pair continues, but will eventually decay into 2 Basic Pairs

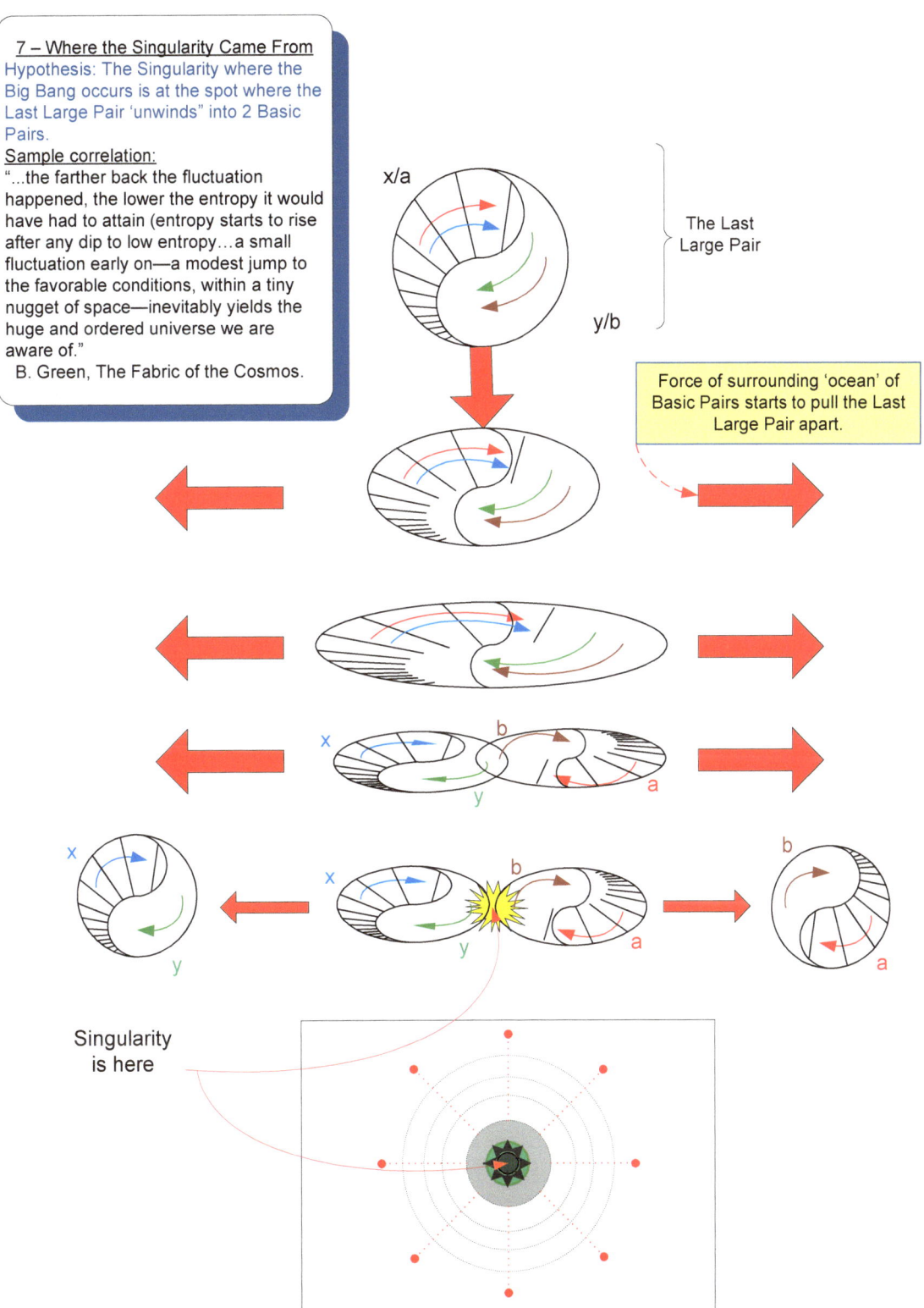

Appendix 2 – Other Predictions and Comments based on the Codex 5 Model

1) String Theory is incorrect

The Codex 5 Model does not need or include any type of String Theory, and it indicates that String Theory is an incorrect way of describing the aspects of physical reality that it refers to.

The Codex 5 Model indicates that as entities of matter/energy become smaller and smaller, they becomes physically simpler and simpler. Hence, the fundamental entity is a Basic Pair (BP), which really has only two key "parts": Absolute Nil and Non-Nil. The BP also has a simple property, described simply as "how it rotates," though this could be just termed "the basic property that explains how any two entities interact." "Rotation" and the accompanying diagrams(above in this document) are really a short-hand way of saying:
a) each entity has a simple property that makes it similar to or different from other entities, and
b) when two entities with different or the same properties interact, here's what happens.

These entities and the simple rules outlined in the Codex 5 Model are all that is needed to "grow" an entire Universe.

If the entire Universe is considered, it is obvious that matter/energy starts as very simple entities (e.g., the Hydrogen atom), and then – through interactions with like entities and/or some force(s) – more and more complex entities evolve.

For example, matter moves from an atom with one proton and one electron, up through atoms with over 100 protons, neutrons and electrons. And of course, atoms combine to form molecules, which are even more complex, and molecules ultimately combine to form plants and animals, which are the most multi-faceted known entities in the universe (e.g., the human brain). In general, in the Universe we see a progression from fairly simple constructions to very elaborate, complex constructions.

But String Theory indicates that the smallest entities in the Universe are actually extremely complex. The mathematics that are said to support String Theory are not only non-intuitive, but are extremely complex. For example, the 4-dimensional Space-Time continuum does not support String Theory, which requires 11 dimensions. This is not to say that the math is wrong because of its complexity; but applying such complexity to what I believe are simple physical entities is a misplaced usage.

As noted elsewhere in this paper, the Codex 5 Model takes a different view, which is best described as a Dynamical System or a Cellular Automata approach. In summary: Start with very simple entities, apply rules, and watch what happens at each iteration of the rules. The rules in the Model show a system (i.e., the Universe) where the simplest of entities (Basic Pairs) interact based on certain rules, and eventually evolve into the actual Universe we live in now.

To refer again to Stephen Wolfram's, *A New Kind of Science,* he wrote: "…it also means that if one once discovers a rule that reproduces sufficiently many features of the universe, then it becomes extremely likely that this rule is indeed the final and correct one for the whole universe." (page 469) And: "To find the behavior of the universe one potentially needs to know not only its rule but also its initial condition. Like the rule, I suspect that the initial conditions will turn out to be simple." (page 1026)

The Codex 5 Model's approach is similar to other Cellular Automata and Dynamical Systems:
- The Model maintains that the Universe started with simple conditions (i.e., simple entities, Basic Pairs).
- It applies simple rules.
- It predicts what WMAP has measured as key features of the current Universe, e.g., the creation and current percentages of the Universe's Energy/Matter that are made of Dark Energy, Dark Matter, and Baryonic Matter.

2) There are no Multiverse, no Additional Universes, no Branes

There are not multiple universes, and the Big Bang wasn't a result of two "Branes" colliding. The Codex 5 Model does not use or need any type of Brane or any additional universes. These are interesting ideas, but like String Theory, they add complexity where none is needed.

Even before William of Ockham, Ptolemy is said to have stated, "We consider it a good principle to explain the phenomena by the simplest hypothesis possible." (The Science of Conjecture: Evidence and Probability before Pascal by James Franklin The Johns Hopkins University Press [2001]. Chap 9. p. 241) Ockham's Razor might be paraphrased as:

> When there are several hypotheses trying to explain something logically, the simplest hypothesis is usually the most accurate (i.e., the hypothesis with the fewest variables, assumptions and other components of an argument).

The Codex 5 Model is much simpler and "cleaner" than theories that require the addition of phenomena that have never been detected anywhere and for which there is no evidence. The proposals of additional universes, entities called "branes," and additional dimensions have no more support than any other result of an active (if misguided) imagination. Creativity and imagination can be important in developing new ideas and theories, but they usually prove to be most accurate when they take off from some credible foundation, rather than depending on day-dreams alone.

****

From time to time, various people state that they don't believe that Dark Matter and/or Dark Energy exist. Items 3 and 4 below contend that they do exist, and provide a few examples of solid scientific observations that support their existence.

****

3) Dark Matter DOES Exist

The Codex 5 Model states that Dark Energy and Dark Matter played a key role in the ultimate evolution of Baryonic Matter, and continue to exist in the Universe today. The existence of Dark Matter is supported by observations such as:

- WMAP: See references elsewhere in this paper.
- Hubble website: Report dated Nov. 11. 2010: "Astronomers using NASA's Hubble Space Telescope received a boost from a cosmic magnifying glass to construct one of the sharpest maps of dark matter in the universe. They used Hubble's Advanced Camera for Surveys to chart the invisible matter in the massive galaxy cluster Abell 1689, located 2.2 billion light-years away."
- The Bullet Galaxy collision: NASA observations directly support the existence of Dark matter: See NASA website, http://www.nasa.gov/vision/universe/starsgalaxies/dark_matter_proven.html. "…a team of scientists working with NASA's Chandra X-ray Observatory has found direct evidence that dark matter is as real as the rings around Saturn… Dark matter revealed itself when the team tried a technique called 'gravitational lensing.'… An unseen force, substance or object had escaped the clouds along with the galaxies and was helping to bend more light. For the first time in history, astronomers caught dark matter at work. 'These results prove that dark matter exists,' declared Clowe."
- Musket Ball Cluster: The NASA website reports: "No evidence is reported for [Dark Matter] self-interaction in the Musket Ball Cluster, consistent with the results for the Bullet Cluster and the other similar clusters."
(see http://www.nasa.gov/mission_pages/chandra/multimedia/musketball.html)

This observation agrees with the Codex 5 Model's prediction that Dark Matter [and Dark Energy] no longer interact after the initial formation of LgPs, VLQs, Octets and XL-12's. The Model goes on to predict that there will be no further interaction among Dark Energy and Matter until the next Big Bang occurs.

4) Dark Energy DOES Exist

Many recent studies have indicated strongly that Dark Energy does in fact exist. The purpose of this paper is not to list all such studies, but here are a few examples:

- Ref "The significance of the integrated Sachs-Wolfe [ISW] effect revisited", T. Ginnantonio, R. Crittenden, R. Nichol, A. Ross, Monthly Notices of the Royal Astronomical Society [September 2012]: This 2-year study "concluded that the likelihood of the existence of dark energy stands at 99.996 percent."
This paper states: "We have shown that our correlation data remain robust with the latest WMAP7 release of CMB data," i.e., like the Codex 5 Model, this study aligns with WMAP findings.

The paper concludes: "It is clear that, if the ΛCDM model is the true underlying model of cosmology, the significance of the ISW effect will remain lower than some other cosmological probes; however, it represents nonetheless a unique signal which allows us to independently confirm the presence of dark energy through its impact on structure growth and potentially detect deviations in how gravity works to build cosmic structures."

- Shirley Ho; Hirata; Nikhil Padmanabhan; Uros Seljak; Neta Bahcall (2008). "Correlation of CMB with large-scale structure: I. ISW Tomography and Cosmological Implications". Phys. Rev. D 78: This paper also supports the Integrated Sachs-Wolfe effect (ISW), which is seen as direct evidence of dark energy.

- The Dark Energy Survey began in August 2013. An official press release states: "The survey's goal is to find out why the expansion of the universe is speeding up, instead of slowing down due to gravity, and to probe the mystery of dark energy, the force believed to be causing that acceleration." The Codex 5 Model predicts that this Survey will corroborate WMAP data, e.g. that Dark Energy makes up about 71.4% of the energy/mass in the Universe. It will be interesting to monitor the DES' findings during the next few years.

www.ingramcontent.com/pod-product-compliance
Lightning Source LLC
Chambersburg PA
CBHW051025180526
45172CB00002B/472